Biogas in the 21st Century - Developments and Perspectives

Edited by Lina Montuori and Manuel Alcázar-Ortega

Published in London, United Kingdom

Biogas in the 21st Century - Developments and Perspectives
http://dx.doi.org/10.5772/intechopen.1003404
Edited by Lina Montuori and Manuel Alcázar-Ortega

Contributors
Abdelhafidh Benreguieg, Alelign Anmaw Getahun, Guillermo Yorel Noriega Aquise, John O'Connell, Kai Heinlein, Kofi Ofori, Lina Montuori, Manuel Alcázar-Ortega, Melaku Esubalew Cherie, Rosemarie Wagner, Surafel Argahegn Agdew, Tesfa Nega Gesese, Wassihun Guade Alemu, Wilson Hago

First published in London, United Kingdom, 2025 by IntechOpen
IntechOpen is the global imprint of INTECHOPEN LIMITED, registered in England and Wales, registration number: 11086078, 167-169 Great Portland Street, London, W1W 5PF, United Kingdom

For EU product safety concerns: IN TECH d.o.o., Prolaz Marije Krucifikse Kozulić 3, 51000 Rijeka, Croatia, info@intechopen.com or visit our website at intechopen.com.

British Library Cataloguing-in-Publication Data
A catalogue record for this book is available from the British Library

Biogas in the 21st Century - Developments and Perspectives
Edited by Lina Montuori and Manuel Alcázar-Ortega
p. cm.
Print ISBN 978-1-83634-266-3
Online ISBN 978-1-83634-265-6
eBook (PDF) ISBN 978-1-83634-267-0

If disposing of this product, please recycle the paper responsibly.

Meet the editors

Lina Montuori obtained her Ph.D. cum laude in Industrial Engineering and MBA from the Polytechnic University of Valencia (Spain). Previously, she had received her B.Sc. and M.Sc. cum laude in Mechanical Engineering from the University of Naples "Federico II" (Italy) and Executive Master in Digital Marketing and Analytics from the Catholic University of Avila (Spain). She worked at the Power Center for Utility Explorations at the State University of New York (SUNY) in Buffalo (USA). Furthermore, she has a demonstrated track record of successfully implementing industrial projects at international top-tier companies, including Enel, Solaer Group, and Toshiba Mitsubishi-Electric Industrial Corporation. Currently, she is a lecturer at the Department of Applied Thermodynamics of the Polytechnic University of Valencia. Furthermore, she is a senior researcher at the Energy Engineering Institute (IIE) with multiple and relevant publications in high-impact journals in renewables, smart grids, energy management, biofuels, and port automation.

Manuel Alcázar-Ortega received his Ph.D. in Industrial Engineering from the Polytechnic University of Valencia (Spain) and a Ph.D. in Electrical Engineering from the University of South Florida (USA). Previously, he had received his B.Sc. and M.Sc. in Industrial Engineering from the University of Castile-La Mancha (Spain) and the Polytechnic University of Valencia, respectively. Currently, he is an Associate Professor and Deputy Director of the Department of Electrical Engineering at the Polytechnic University of Valencia, where he has been linked to the Research University Institute for Energy Engineering since 2004. There, he conducted research activities through various international projects in collaboration with network operators, large and medium-sized energy consumers, distributors, energy traders, and regulators in Europe, the United States, and other South American countries.

Contents

Preface

In the face of global population growth and unpredictable climate change, biogas is gaining momentum as a renewable energy source for multipurpose applications, including heating, electricity generation, and direct use as fuel. At the global, national, and local levels, biogas plays a pivotal role in countries' sustainable development, contributing to reducing ineffective waste, mitigating negative environmental impacts, and creating economic opportunities.

In this framework, the present book aims to provide an overview of the potential of biogas as a sustainable, affordable, and low-emission fuel that can convert waste into energy. Unlike fossil fuels, biogas can help secure the country's energy independence in the face of external political uncertainties.

The chapters within this volume address key thematic areas, beginning with the current status of biogas production, the limitations of existing technologies, and exploring sustainable strategies for the development of biogas-based energy communities as a means of reducing energy dependency on a large scale. This includes mitigation techniques, regulatory frameworks, and their broader climate implications. Moreover, this discussion is complemented by in-depth examinations of third-generation biofuels derived from marine organisms as a promising alternative to carbon-based fuels, offering potential and unique advantages, as well as emerging technologies in hydrogen production from biogas.

The authors have reviewed the current status of biogas production to outline the opportunities and challenges associated with various biogas conversion technologies, including anaerobic digestion and gasification processes. Furthermore, the contribution of biogas to meaningful decarbonization throughout the entire energy production supply chain has been addressed, along with its relevance in promoting a circular economy for waste management, thereby enabling the optimal use of the several billion tons of biological waste produced worldwide each year.

Additionally, a novel perspective on how policy can support biogas integration with other renewable energies has been provided by analyzing key operating parameters, technology limitations and economic constraints. Among the various production technologies, a specific book chapter has been dedicated to estimating the positive impact of compact biodigesters on the optimization of thermal comfort in the residential sector, thereby enhancing the modern concept of bioclimatic and sustainable homes. An additional section of the book delves into membrane cover technologies, assessing the multiple advantages they own as biogas storage systems in terms of flexible storage, cost-effectiveness, and chemical resistance.

Additionally, the authors have conducted an extensive analysis of the potential of third-generation biofuels from algae, including both microalgae and seaweeds, highlighting the unique and beneficial properties of aquatic photosynthetic organisms, such as high CO_2 sequestration capabilities, high biomass yields, and rapid growth rates.

Subsequently, emphasis has been placed on contemporary technologies for hydrogen production from biomass waste. The analysis of economic, environmental, and societal implications of using biogas for H2 production offers the reader an interesting prospect about the available pathways and possible barriers to its sustainable production.

Finally, the book concludes by presenting the risks of forming hydrogen sulphide (H_2S) during biogas production and how this impurity can inhibit methanogenesis and lead to geosequestration.

We believe this book will serve as a valuable resource for scholars, stakeholders, policymakers, and professionals interested in gaining a deeper understanding of the current state and future directions of biogas production and its applications in the 21st century.

Finally, we sincerely thank all contributing authors and editors who have diligently worked to create this comprehensive volume.

Lina Montuori and Manuel Alcázar-Ortega
Universitat Politècnica de València (UPV),
Valencia, Spain

Chapter 1

Perspective Chapter: Biogas Production – Current Status, Challenges, and Future Perspectives

Surafel Argahegn Agdew, Melaku Esubalew Cherie, Wassihun Guade Alemu, Tesfa Nega Gesese and Alelign Anmaw Getahun

Abstract

An estimated several billion tons of biological waste are produced each year as a result of the world's population growth and population density. The conversion of garbage to biogas that is a prime example of waste-to-energy conversion is a great area of interest due to the increasing demand for renewable energy sources. This approach, which addresses two problems simultaneously (conversion of waste and production of biogas), is regarded as one of the best alternative energy sources. Biogas, produced through anaerobic digestion, is clean, affordable, sustainable, and environmentally friendly. This chapter assesses the current status of biogas production, its challenges, and future directions. In this work, all substrates for biogas production and treatment methods with its state-of-the-art and current applications were reviewed. The current status of anaerobic digestion technology and its challenges were also addressed. And finally, the overall challenges such as technological limitations, economic constraints, and environmental concerns, and future directions for biogas production, including advancements in technology, policy support, and integration with other renewable energy sources to enhance sustainability and economic viability were all addressed.

Keywords: biogas, anaerobic digestion, substrate, renewable energy, waste

1. Introduction

Based on the widely documented correlation among energy access, economic growth, and living standards, the trajectory of world history and development is intricately connected to energy production [1]. The energy resources currently used are generally categorized as renewable or non-renewable [2]. Non-renewable resources consist of natural gas, oil, and coal, whereas renewable resources include biomass, solar energy, water, and wind [3–5]. Early civilizations used renewable resources for energy, but since the industrial revolution and the creation of the steam engine in the eighteenth century, fossil fuels have surpassed renewables [6]. Currently, the increasing energy request driven by population and industrial growth continues to be largely met by energy resulting from fossil fuels [7]. The excessive use of fossil fuels leads to the release of significant amounts of greenhouse gases (GHGs) into the atmosphere, contributing to global warming and climate change [7–9]. Additionally, the declining fossil fuel supplies will lead to higher costs for these resources. Due to this, there is a pressing need to find alternative energy sources, with a particular emphasis on renewables. Because of its carbon neutrality, low amount of pollution emissions, and its potential to increase energy security, renewable energy is beneficial [10].

1.1 The world energy requirements

The world's energy demand is rising at a never-before-seen rate due to population expansion and industrialization [11]. Even though the field of renewable energy technology is developing quickly these days, fossil fuels still control most of the energy industry. Global energy demand has grown exponentially over time, with an estimated 85% increase between 2010 and 2040. Non-renewable fossil fuels including coal, oil, and natural gas account for around 85% of the world's energy supply. These fuels produce considerable greenhouse gas (GHG) emissions, particularly CO_2. The ongoing usage of fossil fuels is causing environmental degradation due to increasing CO_2 emissions and raising the long-term risk of energy insecurity. Currently, several countries in the world have faced huge economic losses due to climate change and ecosystem destruction through the utilization of fossil fuels as well as reckless resource development [12]. Thus, many governments are actively financing the development of alternative energy sources, and different related studies have been carried out to switch to renewable sources. Due to the depletion of fossil fuel reserves, which account for 84% of the world's energy consumption, promoting the use of renewable energy has become an essential global effort [13]. Today, the use of renewable energy sources is growing, and efforts are being made to advance current technologies [3]. Renewable energy sources will account for 27% of all energy production by 2030, according to the European Union's (EU) renewable energy plan [14].

There are notable differences in the biogas sector across the globe. Various nations have developed a variety of biogas systems, primarily based on varying conditions, energy supply chains, and demand. While the breakdown of sewage is the predominant method of producing biogas in Switzerland and Sweden, the UK, Australia, and South Korea used landfill sites to attain a significant amount of their produced biogas. Because manure is readily available and abundant, it is mostly used in Denmark. The majority of biogas produced in Germany, the UK, and Sweden comes from food waste [15].

China and Germany are acknowledged as the world leaders in farm-based biogas production since China has over 24,000 small-scale plants and Germany has almost 8000 agricultural plants. In a similar vein, significant farm-based biogas plants were used in France, Holland, Austria, and Italy. Furthermore, the size of plants varies from tiny home units to larger plants that employ manure, industrial waste, and domestic waste as feedstocks to produce both heat and power. The majority of installed biogas plants in Asia and Africa were family-sized, according to studies [7]. The global market for microscale biogas has been controlled by China and India. Thailand currently benefits from over 150 industrial waste plants and over 1700 biogas plants. The Thailand's government has made an effort to develop industrial wastewater technology, which could produce 7800 TJ of biogas annually [16]. The majority of the villages, approximately 2800 out of 3915 in all 75 districts of Nepal, have small-scale or domestic biogas production systems, according to the Ministry of Energy of Nepal. In Nepal, two main plant classifications have been established. These are the Indian-style floating-drum plants and fixed-dome plants, which have a concrete-prepared dome, a cylindrical digester, and a flat floor [17]. Of the 50 million microscale digesters in use worldwide, 42 million are situated in China, and an additional 4.9 million are in India. Only 700,000 biogas plants are operational in Asia, Africa, and South America, according to data from the World Biogas Association [18].

Approximately 7000 large-scale biogas systems are now in operation in China. In 2017, Europe accounted for 17,783 plants, with Germany leading the European biogas market with 10,971 plants, followed by Italy (1665), France (742), Switzerland (632), and the UK (613), in that order (Association 2018). According to data from the World Biogas Association, there are over 2200 large-scale anaerobic digester plants in the USA that can produce 977 MW. Biogas can also be upgraded to biomethane. Despite being a relatively new procedure, it has gained extensive adoption globally. Some biogas upgrade plants create car fuel, while others deliver it to the local or national grids association [18].

Although it has made only modest strides in the field, Africa boasts a wealth of varied resources for the generation of biogas [19]. Profitable biodigesters still need to be developed, despite the continent's notable progress in small-scale biogas facilities [19]. Potential employers of large-scale biogas technology in Africa include livestock and harvest farmers, small- to medium-sized and large food treatment companies, wastewater, sanitation, and municipality operating institutes, as well as municipal waste management organizations. Additionally, biogas facilities and technology have the potential to help hospitals, schools, universities, and commercial structures [20]. The technological advancement of the commercial biogas system in Africa has not been adequately documented in the scientific literature, with the exception of South Africa. The lagoon, plug low, and upflow sludge blanket (UASB) are developed technologies in South Africa [21]. Around 15 terawatts (TW) of energy is consumed annually worldwide. Only 7.8% (0.94 TW) of this comes from renewable sources [22]. Only 4% of the world's energy consumption occurs in sub-Saharan Africa, which is extremely small given that 13% of the world's population lives there [23]. Fuel wood accounts for about half of Africa's total energy use. Numerous drawbacks are linked to this fuel type, such as the accumulation of indoor smoke and its detrimental effects on health, deforestation, and the labor-intensive and hazardous procedure of gathering wood [18]. As a result, alternative renewable energy sources are essential and using biogas as an energy source is one viable substitute [16].

1.2 Biogas: A viable renewable energy source

Biogas is produced from the anaerobic digestion (AD) of organic materials and comprises mostly of flammable methane (60–70%), which is the chosen gas for energy production, carbon dioxide (30–40%), and trace amounts of other gases, such as hydrogen sulfide [24]. Biogas is generated in biogas plants through the anaerobic bacterial decomposition of biomass. Biomass is categorized into three types: (1) farm-based substrates like liquid manure, food wastes, harvest wastes, and energy crops; (2) organic wastes from households and municipalities, including separately collected organic wastes, market wastes, and expired foods; and

(3) industrial by-products [25] such as glycerin, by-products of food processing, or waste from fat separators [11, 13, 14, 23]. The production of biogas through anaerobic digestion involves a series of stages: hydrolysis, acidogenesis, acetogenesis, and methanogenesis as shown in **Figure 1**. In hydrolysis, complex organic matter is broken down into simpler compounds by hydrolytic bacteria. Acidogenic bacteria then convert these into volatile fatty acids during acidogenesis, producing byproducts such as ammonia and carbon dioxide. In the acetogenesis stage, acetogenic bacteria further transform these fatty acids into acetic acid and other gases [26]. Finally, methanogenic archaea convert acetic acid, carbon dioxide, and hydrogen into methane, which constitutes the primary component of biogas, all occurring in an oxygen-free environment. The bacteria are similar to those found in the stomachs of ruminants. Energy production and waste management have various environmental implications [27].

To provide a sustainable solution for the increasing amount of waste generated by a growing global population, an integrated approach must be developed [14]. Biogas technology makes the use of some biomasses to partially meet energy needs quite attractive. Effective biogas systems may save resources and safeguard the environment in a variety of ways for both users and the community [28]. But what distinguishes biogas from other renewable energy sources is its importance in controlling and collecting organic waste while also producing fertilizer and water for agricultural irrigation. Biogas is neither monopolistic nor complex like other renewable energy sources, and it does not require any technology or geographic limitations to generate energy [28]. By using waste for biogas production, greenhouse gas emissions are reduced, heat and energy are produced, and odors from landfill natural decomposition are eliminated. Additionally, digestate byproduct of anaerobic digestion can be utilized as a naturally occurring, nutrient-rich fertilizer. Thus, the process of producing biogas has several advantages that can help with a variety of global issues related to waste management, sanitation, socioeconomic issues, sustainable development, and environmental preservation [11, 27].

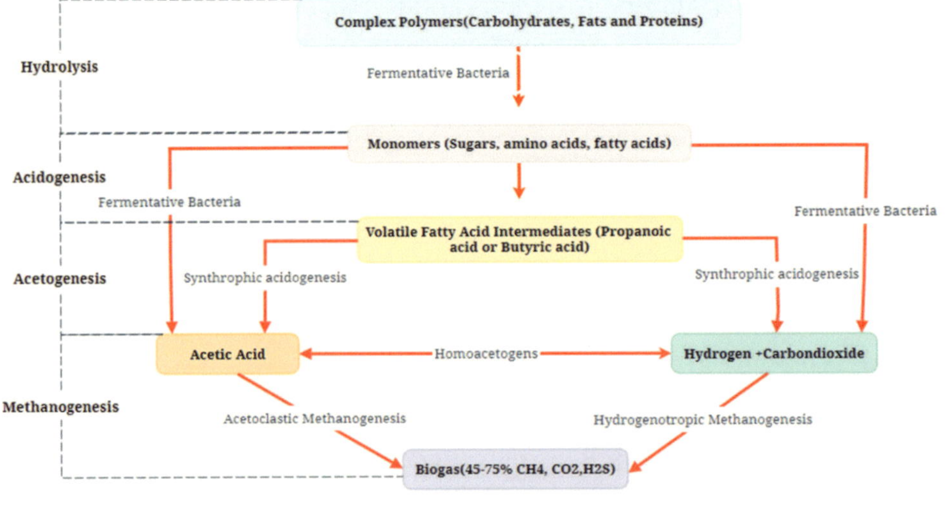

Figure 1.
Stages of biogas production [18].

1.3 Substrate assessment for biogas production

The initial biogas facilities were based on corn silage and liquid co-substrates from processing and agriculture (manure), and they were modeled after German technologies [29]. By-products from the agri-food sector gradually took the role of silage from the targeted crops. Numerous feedstocks that can be utilized to manufacture biogas are produced by agriculture and industries that rely on agricultural components. Plant and animal products that contain organic components and have the ability to ferment are suitable substrates for biogas plants [25]. There are two types of raw materials used in the manufacture of biogas: mono- and co-substrates. Mono-substrates, such as slurry, manure, and the stomach contents of animals, particularly ruminants, can ferment because they contain methane bacteria. They also contain a wide range of macro- and micronutrients that are essential for the development of microorganisms. To improve process efficiency, attain proper hydration, or avoid inhibition, co-substrates are introduced to the digester. These consist of greenhouse waste, energy crop-focused biomass, agricultural wastes, and byproducts [30]. The biggest percentage of substrate utilized for agricultural biogas production in 2021 was made up of distillery stock (18.98%), manure (16.41%), agro-food industry waste (16.61%), and leftovers from fruit and vegetable processing (16.20%) (fruit and vegetable residues, post-production waste) [23]. The main factors contributing to these raw materials' popularity are their wide availability, simplicity of transportation, and low acquisition prices. Straw, grasses from permanent grasslands that have not been farmed, or biomass that can be pulled out from marginal land that is not appropriate for growing food continue to be important and neglected resources [31]. These feedstocks are difficult to ferment without the application of suitable pretreatment techniques [32]. Agro-energy is increasingly being cited for growing global food prices, and this feature is prompting the introduction of second- and third-generation biofuels, including those derived from waste, lignocellulosic feedstocks, and algae [33]. Without lowering the amount of land resources required to ensure the nation's food security, Polish biogas plants will produce more biogas by using more agricultural and processing wastes and byproducts. Simultaneously, the growth of second-generation biofuel production, sensible waste and by-product management, and the integration of the biogas sector with the concept of a closed-loop economy will be observed. But this calls for the assistance of contemporary methane fermentation technologies as well as techniques for preconditioning a novel variety of substrates [22, 28, 30, 33–35].

2. Methodology

This study employed an extensive literature review methodology to investigate the current status, difficulties, and potential of biogas generation in the future. The use of feedstock, anaerobic digestion, biogas technology, and renewable energy policies were all the subjects of a comprehensive collection and analysis of the published research articles, review papers, technical reports, and policy documents. Relevant material was found by searching scientific databases such as Scopus, Web of Science, and Google Scholar, as well as relevant organizational websites such as The World Biogas Association and national energy agencies. The following search terms were used: "biogas," "anaerobic digestion," "biomethane," "renewable energy," "waste-to-energy," "biogas substrates," "biogas technology," "biogas challenges," "biogas policy," "biogas generation status today," "biogas generation prospects for the future," and combinations of these terms.

Current status of biogas production: This required examining data and patterns related to feedstock diversity, anaerobic digestion and biogas upgrading, technological advancements, global and regional biogas production capacity, and the impact of current regulatory frameworks. To give a quantitative picture of the existing situation, reports and statistical data on the installation of biogas plants and energy production were analyzed.

Challenges of biogas production and utilization: The main challenges to the broad usage and effective functioning of biogas technology were identified and categorized in this section. These difficulties were examined in terms of technical and infrastructure restrictions, financial limitations, digestate heavy metal contamination concerns, building and running expenses, the need for biogas upgrading and purification, and political and policy obstacles.

Future perspectives of biogas production: Examining possible future paths and prospects for the biogas sector was part of this component. This involved examining new developments in technology, the function of incentives and support from policy, the possibility of integrating biogas production with other renewable energy sources, and methods for improving the economic feasibility and sustainability of biogas production. The collected data was then organized to offer a thorough summary of the production of biogas as it is today, a critical evaluation of the difficulties encountered, and a perceptive outlook on the direction this renewable energy source may take in the future. Key trends, knowledge gaps, and possible areas for further study and development in the field of biogas generation were identified using the results of the literature review.

3. Current status of biogas production

The generation of biogas has drawn a lot of interest as a renewable energy source, especially in light of the growing need for energy worldwide and the requirement for environmentally friendly waste management techniques. The global trend in biogas generation is increasing driven by the desire for cleaner energy sources, better waste management techniques, and supportive government policies [17]. This overview presents recent accomplishments in the field of biogas production, including global capacity and growth, technological advancements in biogas upgrading, legislative frameworks enabling biogas growth, environmental impact, and sustainable benefits.

3.1 Global capacity and growth

The capacity to produce biogas worldwide has grown significantly, rising from around 65 GW in 2010 to over 120 GW in 2019. The demand for renewable energy sources and increased awareness of climate change are credited with this expansion. With the large investments in biogas facilities and technologies, Europe continues to lead the world in biogas generation, accounting for over 70% of total output [18]. **Figure 2** shows biogas production capacity in (MW). Europe and European Union take the largest share, while African nations and South America have the lowest biogas production capacity.

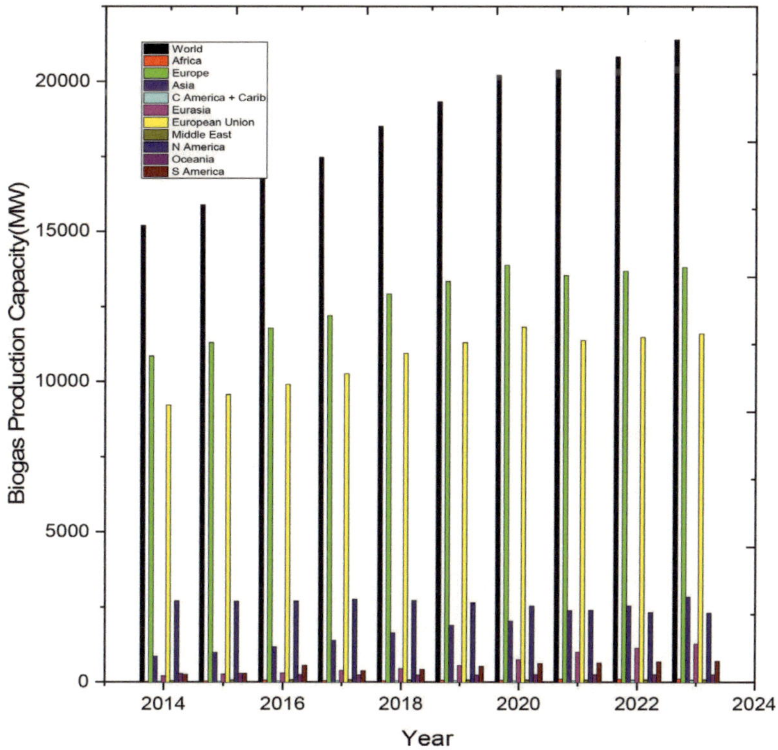

Figure 2.
Biogas production capacity (2014–2023) [17].

3.2 Feedstock diversity

The primary feedstocks for the biogas plant are organic materials, which can be classified into many types. Research has been done to assess the possibilities of producing biogas from organic materials found in rural areas. The amount of feedstock in biogas plants may be reduced by the highly fermentative wastes [25].

Microalgae with favorable characteristics could be used as feedstock in biogas systems. Microalgae is a versatile option for biogas plants because of its increased lipid and carbohydrate content, improved efficiency, and easier production when compared to other biomass resources [33]. Animal manure, food waste, municipal solid waste, and agricultural residues are just a few of the organic elements that can be used to create biogas. The feedstock selection greatly impacts the yield and production efficiency of biogas. For example, lignocellulosic materials frequently need to be pretreated to improve digestion, whereas food waste and garden waste are recognized for their high biogas generation per unit of fresh weight [16]. **Figure 3** shows the current feedstocks for biogas production and generations.

3.3 Technological advancements

Improved anaerobic digestion procedures and biogas upgrading methods are examples of recent advancements in biogas production technologies. To produce high-quality biomethane appropriate for grid injection or automobile fuel, these developments concentrate on increasing methane yield and lowering contaminants

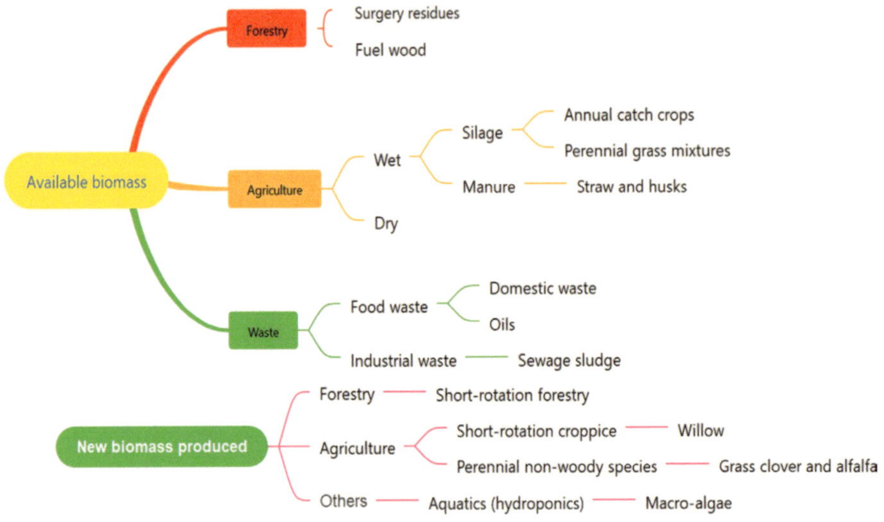

Figure 3.
Substrate classifications for biogas production [29].

such as CO_2 and H_2S. Furthermore, the optimization of microbial consortia for more effective biogas production is being investigated through the integration of molecular biology approaches [7].

3.3.1 Physicochemical technologies

Various technologies are being utilized for the treatment and pre-treatment of biogas and feedstock, along with its enhancement, as shown in **Figures 4** and **5**. Pre-treatment methods encompass mechanical, microwave irradiation, extrusion, liquid hot water, steam explosion, and chemical (alkaline, acidic, organosolv) and oxidative techniques that enhance biogas production from feedstock.

The biochemical advances for improvement of biogas generation handle incorporate fungal, microorganism, and enzymatic pre-treatment are modern and beneath investigate. These advances are greener, eco-friendlier, and appropriate to cultivate the execution handle. For the improved biogas feedstock handle of enzymatic

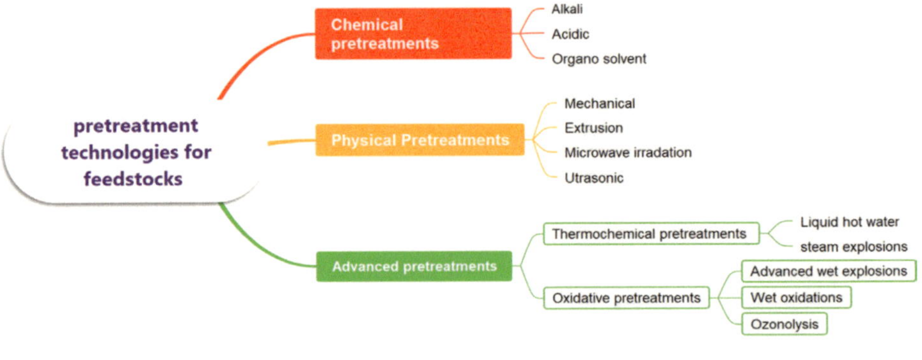

Figure 4.
Classification of different advanced technologies for feedstock treatment used for biogas production [7].

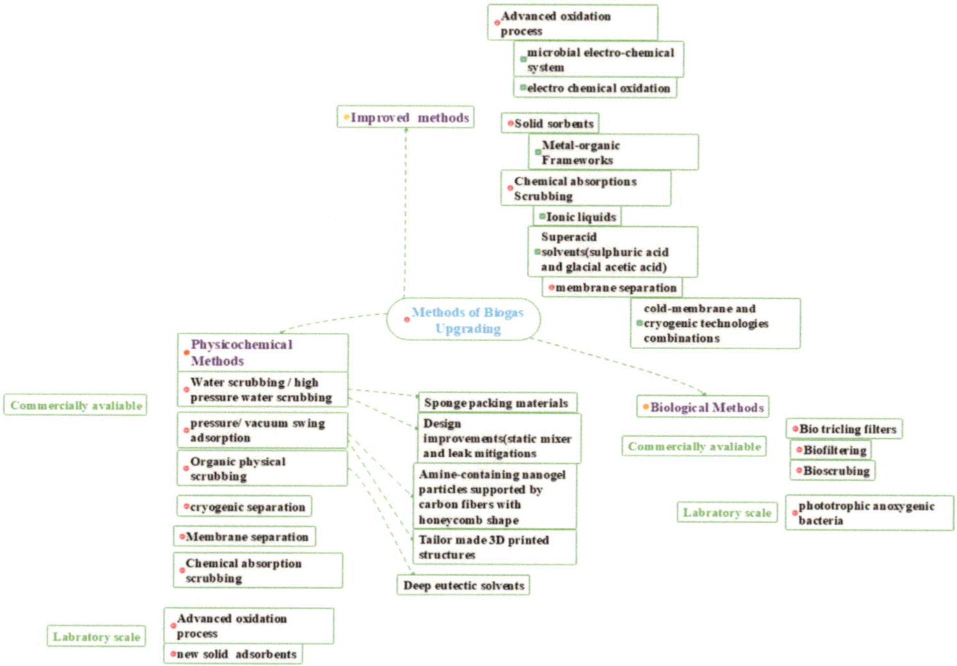

Figure 5.
Summary of biological and physicochemical methods for biogas upgrading as well as their possible new improvements [36].

hydrolysis amid the anaerobic assimilation prepare, it is worth to develop the method of microbial development on utilized biomass, which might be done by biochemical advances. Different enzymes such as xylanases, proteases, cellulases, and ligninolytic chemicals have been utilized for the transformation of feedstocks in an anaerobic assimilation prepare into edible sugars and to progress generation of biogas rate beneath chosen operational parameters to dodge inhibitory compound arrangement. In common, for the treatment of ligno-cellulose biomass other advances must be utilized some time recently enzymatic liquefaction for getting improved generation rate of sugars and surrender.

The primary purpose of upgrading technologies is to remove impurities such as hydrogen sulfide (H_2S), water vapor, ammonia, and siloxanes, which can adversely affect the usability of biogas, its compatibility with natural gas, and its acceptance by end users. A secondary objective is to extract carbon dioxide (CO_2) to enhance the calorific value of biogas and reduce its density, thereby ensuring it meets the specific standards of the Wobbe index during the upgrading process. Furthermore, depending on the technology utilized, the storage, application, and removal of CO_2 can transform biogas into a carbon-neutral energy resource, effectively managing anthropogenic CO_2 emissions. The final product of the biogas upgrading process is referred to as biomethane with its quality and performance determined by its intended use. Typically, biomethane comprises 95–99% methane, 1–6% carbon dioxide, and 0.02–0.05% hydrogen sulfide. Technologies for upgrading biogas include scrubbing, membrane separation, pressure swing adsorption, and cryogenic methods. The scrubbing process can involve chemical and organic solvents or physical scrubbing techniques [9]. Over the last decade, the European bio-methane production has increased by 540% and its trend keeps growing at a constant rate, reaching an energy equivalent of 32 TWh in 2020 (23% higher than that in 2019). According to the EBA database, it is expected that 30 to 40% of the total gas consumption in Europe will be covered by bio-methane by 2050. In this context, the deployment and development of innovative, sustainable, and cost-efficient (bio) technologies for biogas upgrading is crucial toward achieving a mature bio-methane industry. Despite the clear potential of AD-derived biogas as a platform to produce renewable energy from organic waste sources, the economic added value of biogas produced *via* AD is limited [5, 6], unless otherwise, it has to be upgraded with appropriate technology. The upgrading of biogas, which is essential for its use as fuel, mostly uses physicochemical technologies to extract CO_2 and eliminate contaminants such as H_2S. By employing materials such as activated carbon, pressure/vacuum swing adsorption (PSA/VSA) selectively adsorbs CO_2 under pressure, releasing it upon depressurization or vacuum. Recent developments center on new adsorbents and system architectures, such as 3D-printed structures and temperature swing. Utilizing the greater solubility of CO_2 and H2S in pressured water, water scrubbing is another popular technique that is followed by water regeneration [37]. Research on optimal methods and pretreatment for H_2S removal has been prompted by the fact that, despite being straightforward and economical, it may have reduced energy efficiency and methane losses. Biological techniques including biofiltering, biotrickling, and bioscrubbing provide ecologically acceptable and energy-efficient alternatives to physicochemical methods for H_2S removal. In bioreactors, phototrophic bacteria also exhibit promise [36]. The classification and merits as well as the possible improved methods of biogas upgrading for biomethane production are illustrated in **Figure 5** and **Table 1.**

3.3.2 Bioreactor for biogas production

The integration of anaerobic digestion and membrane separation in anaerobic membrane bioreactors (AnMBRs) has proven to be effective in addressing the challenges of slow-growing methanogenic organisms and complex microbial systems in biogas fermenters. This is because it increases biomass concentration and retention, which leads to higher productivity, resistance to toxins, and easier product separation. AnMBRs have produced notable methane content (up to 90%) and yields (up to 0.36 l CH_4/g COD). Anaerobic digestion has also advanced to the point that multi-stage reactor topologies are being investigated to improve process stability and the conversion of organic matter to methane, albeit some multi-stage systems are still expensive. To better understand and regulate the anaerobic digestion process, recent studies have also made use of molecular biology methods [7]. **Figure 6** illustrates the improved and conventional AnMBRs methods for biogas generation.

3.4 Regulatory frameworks supporting biogas growth

Several organizations and countries, like the European Biogas Association and the Chinese government, have enacted rules and legislation to encourage biogas growth and renewable energy. While there are common guidelines in Europe, various countries use National Renewable Energy Action Plans to customize policy to their own energy markets and biomass sources. China has several energy policies that assist the renewable market, and important government authorities are actively establishing biogas laws [18]. These policies cover a variety of

Technologies	Benefits
Water scrubbing	High efficiency: around 97% CH_4 recovery Both CO_2 and H_2S can be removed at once, Suitable for H_2S concentration less than 300 cm^3/m^3 Capacity is modifiable upon alterations of temperature and pressure Less than 2% CH_4 losses Not overly sensitive to impurities Water is a regeneration of water with ease No sophisticated system or chemical requirements Rapidly scalable
Organic scrubbing	High CH_4 recovery efficiency (>97% CH_4) Less volume of solvent required due to the high solubility of CO_2 Concurrent separation of contaminants including H_2S More efficient than water scrubbing low CH_4 losses Recovery of residual methane is possible by heating
Pressure swing adsorption	Highly efficient CH_4 recovery (96–98% CH_4) H_2S is removed. N_2, O_2, and moisture removal are possible High pressure but regenerative Faster installation and relatively easier operation due to compact setup Suitable for small capacities as well as tolerant to impurities Clean and water-free biomethane produced Less energy requirement Fast adsorbent regeneration
Cryogenic separation	Up to 98% high-purity CH_4 recovery High-purity CO_2 recovery as a by-product which can be used for example as dry ice Minimal additional energy is required to reach liquid biomethane (LBM) production level Very low CH_4 loss Eco-friendly, no chemical usage is involved
Membrane separation	Cheap investment and operation Both H_2S and H_2O are separated Simple compact design, installation, and operation Easy maintenance without any chemical requirements Low energy requirement High reliability Suitable for low gas flowrates No substantial increase in costings for small units CH_4 recovery efficiency of the single-pass module is around 92% CH_4, for multistage units CH_4 purity reaches more than 96% CH_4 Overall environment-friendly process operation
Chemical scrubbing	High-efficiency CH_4 recovery (>99% CH_4) Complete H_2S removal is possible Compression of raw biogas is not mandatory Can operate at low-pressure conditions Low operational cost Easier regeneration of the solvent Larger volume of CO_2 can be dissolved in each unit of chemical solvent per unit of volume compared to H_2O or organic solvents high selectivity for CO_2 with minimal CH_4 losses, less than 0.1%
Biological upgrading	Isolation of both H_2S and CO_2 is possible Enrichment of biomethane Generation of valuable by-products in some cases Environment friendly No chemical requirements Minimal or no heating required No undesirable by-product or end-product development

Table 1.
Merit of actual biogas upgrading technologies [38].

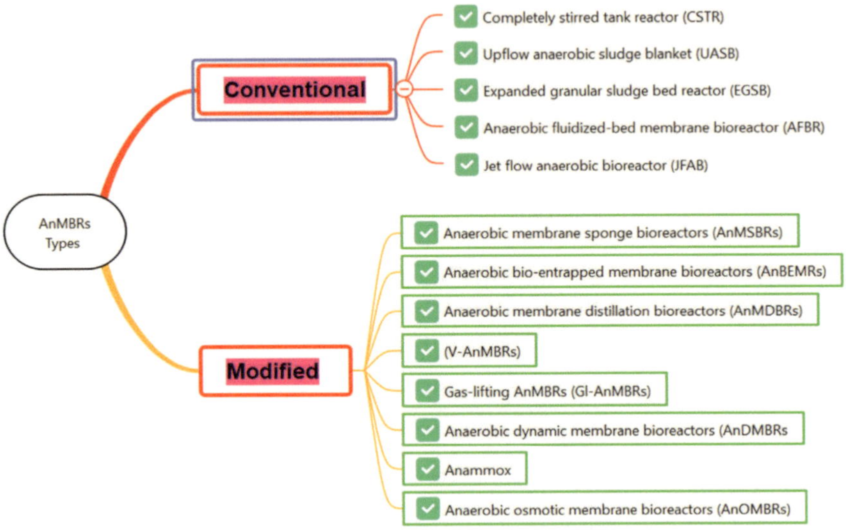

Figure 6.
Improved and conventional AnMBRs methods for biogas generation [7].

issues, including rural biogas construction, waste management, and technology usage, with the goal of encouraging investment, privatization, and cross-sector collaboration, thereby contributing to national priorities and efficient biogas management [22]. For instance, Denmark, Italy, and Germany have 7, 28, and 32 renewable energy programs, respectively, whereas the UK has 118 [18]. Their national plans for arranging renewable energy in Europe are written using this common framework. Several EU countries set up their energy markets and biomass sources in spite of the existence of common European regulations like 2009/28/EC that took into account the production of biogas from agronomic deposits and organic waste and its use in generating power and heat. The National Renewable Energy Action Plan is a set of policies that these nations have adopted to meet their individual requirements and priorities [18].

3.5 Environmental impact and biogas generation

By lowering greenhouse gas emissions and the amount of organic waste that ends up in landfills, biogas generation helps with waste management. The nutrient loop in agricultural systems is closed by the anaerobic digestion process, which also provides digestate, a nutrient-rich fertilizer, in addition to renewable energy [39]. To ensure environmental sustainability, strategic decision-making is essential at every stage of the biogas production lifecycle. According to studies, it is critical to select high-energy density feedstocks, including corn silage and straw, which have smaller land-use footprints and less of an influence on ecosystem diversity, particularly when digesting a single feed-stock. The environmental benefits of using waste materials are further highlighted by co-digestion techniques that concentrate on agriculture and food industry by-products rather than energy crops, which greatly reduce land occupation consequences. Efficiency and environmental implications are significantly increased by increasing the usage of biogas through trigenerational systems for smaller facilities and upgrading biomethane in conjunction with combined heat and power (CHP) for bigger operations, especially with external heat recovery. Crucially, recovering residual biogas from digestate storage reduces strong greenhouse gas emissions and substitutes [40].

4. Challenges of biogas production and utilization

The production and use of biogas presents a viable path toward waste management and the creation of renewable energy. However, a number of obstacles spanning the technological, financial, social, and environmental spheres stand in the way of the broad implementation of biogas technology [41]. To fully realize biogas's potential as a sustainable energy source, these obstacles must be removed.

4.1 Technical and infrastructural challenges

One of the main technological challenges is the shortage of specialized knowledge and technical skills necessary for the efficient operation and upkeep of biogas reactors. This lack of expertise results in regular malfunctions, diminished efficiency, and, in the end, the neglect of biogas facilities. In areas where biogas technology is still emerging the lack of proficient technicians and engineers presents a major hurdle to the ongoing effectiveness of biogas systems [42]. Establishing a sustainable technology is still a significant obstacle to the adoption of biogas in the underdeveloped nations as opposed to their industrialized counterparts. Although biomass is abundant in developing nations, insufficient infrastructure that can support the necessary technology prevents them from making effective use of their resources. Additionally, the success of implementing biogas depends on how well users comprehend the technical basics [43].

Since most biogas operators lack the technical skills to generate biogas, combining biogas technology with eco-agricultural technology has proven difficult [44]. Low biogas output is another effect of it. For example, the feedstock composition (nutrient and organic content, contaminants, and potential inhibitors) affects the amount of biogas generated and can also lead to the process's failure. Furthermore, low production and digester failure can also be attributed to inefficient mixing capacities, a lack of technical competence to build high-quality digesters, and a lack of information about how to properly maintain these digesters. Temperature variations induce process instability in biogas digesters, which negatively impact biogas process performance because microorganisms' activity decreases at temperatures above or below their optimal range [35]. There is no efficient method to control the working temperature in the winter, particularly for colder regions. When the temperature drops below 15°C, digesters without insulators do not work well. Nepal is a common example, with frigid temperatures (about 10°C) in steep areas. There is a scarcity of water in many developing nations. Wet fermentation, a popular technique for producing biogas, uses excessive amounts of water, which is problematic for developing nations that have water scarcity. Furthermore, a large amount of methane residue make it difficult to transport and apply digestate to agricultural areas [17, 21, 44]. Another major technical issue is the restricted availability of vital infrastructure, such as sewage networks and gas pipelines. Biogas distribution and transportation require a robust pipeline infrastructure that is frequently absent in rural or isolated locations. Similar to this, the continuous supply of organic waste to biogas plants depends on the existence of dependable sewage networks. Biogas projects' scalability and financial sustainability may be hampered by the lack of such infrastructure [45].

4.1.1 Heavy metal contamination

In the context of anaerobic digestion (AD), the presence of heavy metals (HMs) accumulated in digestion substrates, including plant residues and livestock manure, can exert a profound influence on biogas yields. Specifically, the presence of Cu^{2+} (ranging from 0 to 100 mg/L), Fe^{2+} (between 50 and 4000 mg/L), Ni^{2+} (from 0.8 to 50 mg/L), Cd^{2+} (0.1 to 0.3 mg/L), and Zn^{2+} (up to 5 mg/kg) can stimulate biogas production. Conversely, elevated concentrations of these metals are inhibitory to AD [46]. Given the complex characteristics of the anaerobic system, heavy metals can engage in various physiochemical processes such as (1) precipitation as sulfides (except for chromium), carbonates, and hydroxides, (2) sorption onto solid substrates, which may consist of either biomass or inert particulate matter, and (3) the creation of complexes in solution with intermediates and byproducts produced during digestion. In addition to the physiochemical forms, differences in substrate types, bacterial genera, and environmental factors contribute to the significant variability (ranging from a few to several hundred mg L^{-1}) noted in both the reported levels of heavy metals and their corresponding toxicity levels [47].

Heavy metals are absorbed through feed during both poultry and livestock breeding processes; however, the majority of them are not used and are excreted immediately through animal metabolism. When manure that contains heavy metals is used to generate biogas, the resulting digestate will still have these toxic substances. It is important to highlight that the presence of heavy metal residues in feces adversely affects biogas production because they can be detrimental to microorganisms if present in excessive amounts [48]. The morphology of each metal during the anaerobic digestion process determined its bioavailability, which was a crucial component in predicting the ecological risk of heavy metals [49]. It was controlled by AD operating parameters such as pH, redox potential, and VFAs concentration [50]. Due to their non-degradability, the heavy metals may build up to potentially harmful levels in the digested form [51]. The environmental and hazardous effects of heavy metals are governed by their solubility and availability [28]. If biogas digestate is utilized for fertilizers or other applications without proper treatment, it may harm soil, water resources, and living organisms. Research has indicated that heavy metals such as Zn, Pb, and Cd found in biogas slurry used for soil irrigation can exceed acceptable limits leading to additional risks for human health through the consumption of vegetables and grains irrigated with it. Currently, numerous countries, including the United States, Canada, the European Union, and Japan, have established specific regulations concerning the maximum allowable levels of heavy metals in digestate, but these regulations still do not completely eliminate the potential ecological risks associated with its use [52].

4.2 Economic challenges

The overall investment for biogas generation and utilization is closely linked to the selection of the technical approach and the reliability of the biogas project. This is associated with subsequent operational efficiency, stability, product quality, and output. Therefore, it is essential to evaluate the advantages and disadvantages prior to making an investment. The recovery period of the investment, the internal benefit period, the asset return rate, and the sales profit margin are key metrics for the economic assessment of an investment project. Typically, the payback period for domestic project construction ranges from 6 to 8 years, and the internal rate of return is generally above 15% before tax in the domestic market [50]. One of the biggest economic obstacles to the construction of biogas plants is the high upfront costs, particularly in developing nations. Reactors, equipment for gas upgrading, and other necessary parts can be too expensive for small-scale farmers and business owners. To get past this obstacle, it is essential to have access to reasonably priced financing and investment capital [41].

Another financial issue is that biogas systems frequently have to contend with competition from less expensive energy and waste treatment sources [16]. Despite its negative environmental effects, landfilling organic waste is still a more cost-effective solution than anaerobic digestion in many areas. In a similar vein, biogas may become less competitive in the energy market if fossil fuels are subsidized. To encourage the use of biogas technology, these economic inequalities must be addressed. Additionally, building a biogas plant requires a substantial capital investment, which can be a major obstacle, especially for smaller operations and agricultural communities. The upfront costs of obtaining funding and navigating intricate permitting procedures increase, postponing project implementation and possibly turning off potential investors. These financial burdens could be reduced and increase involvement in biogas projects promoted by streamlining regulatory frameworks and providing financial incentives such as grants, tax credits, and loan guarantees. Policies that internalize the environmental costs of competing waste management and energy generation methods, such as carbon pricing or landfill taxes, could level the playing field and emphasize the true value proposition of biogas in addition to providing direct financial support. In the end, maximizing the potential of biogas as a sustainable waste management solution requires a multifaceted strategy that tackles the financial and legal obstacles [53].

4.2.1 Construction and operating costs

The expenses for construction encompass the prices for direct construction materials, land rental, transportation, and any losses from maintenance. It also accounts for the costs of employing workers, including basic wages for skilled labor, coolies, and handymen, along with bonuses and allowances, as well as provisions for accommodation and utility expenses. Additionally, costs related to equipment such as fermentation tanks, internal combustion engines, and generators are included. Among these infrastructure expenses, the factory's location would influence land rent, labor costs, and the costs of raw material supply. The operational costs of the plant involve the provision of essential raw materials, electricity, heating, and water necessary for the equipment's operation, regular equipment inspections and maintenance, and environmental safeguards. It is important to note that the availability of energy and raw materials can fluctuate with the seasons, requiring more electric heating to keep the fermentation temperature consistent during winter. Furthermore, autumn and winter tend to see an increase in poisoning accidents, which can lead to casualties [15, 19, 54, 55]. Additionally, the waste and emissions produced by biogas power generation initiatives have led to environmental pollution. From the perspective of minimizing risks, technological advancements have the potential to significantly lower operational expenses; for instance, optimizing equipment, substituting raw materials, enhancing technological processes, implementing improved business management networks, and adopting adaptable management strategies could greatly boost the market competitiveness of products in these operational management areas [53].

4.3 Biogas upgrading and purification

The surge in the world population has resulted in increasing ecological worries because of human-caused greenhouse gas (GHG) emissions and the intensification of harmful industrial, agricultural, and municipal discards. Biogas can resolve these issues by generating energy from wastes, therefore lessening global warming [40]. Besides environmental advantages, there are various economic perks of biogas, such as the production of a high-quality digestate from anaerobic digesters for its use as a fertilizer, or biogas as a potentially inexpensive and readily accessible local source of heat and power in isolated and small communities. The considerable ecological and economic benefits of biogas have led to a remarkable growth in the number of biogas upgrading plants in Europe, increasing from 483 to 729 from 2018 to 2020 [22]. Presently, it is estimated that 20% of worldwide gas demand can be supplied by the complete utilization of available organic wastes toward the production of biomethane. Nevertheless, significant technological barriers exist in present biogas cleaning and upgrading technologies, and only 5% of the biogas produced worldwide is being used in the manufacture of biomethane [2]. Biogas can potentially play a key role in the world's energy supply security and diversification. Biogas as a renewable energy resource can be broadly recognized as a carbon-neutral fuel, which reduces anthropogenic greenhouse gas emissions, mitigates global warming, and diversifies energy supply. However, the biogas share in the global renewable energy supply chain and technology deployment and maturity do not correspond with the potential. The impurities found in raw biogas, which can damage equipment and lower the gas's energy content, include siloxanes, carbon dioxide, and hydrogen sulfide. The processing of biogas is primarily undertaken for two crucial purposes: firstly, to eliminate contaminants that are incompatible with the natural gas network, equipment, or consumers (biogas cleaning); and secondly, to separate out carbon dioxide with the aim of improving the energy content, reducing the specific gravity of the refined biogas, and achieving the required Wobbe Index (biogas upgrading). By using different purification technologies, these impurities must be eliminated in order to upgrade biogas to biomethane, a higher-value fuel. For small-scale biogas plants in particular, these technologies can be costly and energy-intensive. One major obstacle to broader adoption of biogas is its economical purification [9]. Table list outs challenges and benefits of biogas upgrading technology for biomethane productions. The challenges in biogas cleaning are articulated in **Table 2**.

4.4 Policy and political challenges

One major regulatory obstacle is the absence of comprehensive national policies created especially to encourage the development of biogas [1]. Investors and developers may be hesitant to fund biogas projects in the absence of clear policy frameworks because they face uncertainty. To create a supportive regulatory environment, national biogas policies that offer incentives, standards, and guidelines must be established [42]. The vast majority of developing nations lack a dependable mechanism for collecting and sorting waste. Trash managers frequently gather various trash streams and dispose of them in uncontrolled dumps or in open fires [16]. There is no defined policy on the usage of renewable energy, and the environmental policy is extremely weak. Another major obstacle impeding the growth of biogas implementation is the lack of government commitment and the poor continuity of prior biogas program projects through succeeding governments [51]. Corruption is

Technologies	Challenges
Water scrubbing	High investment and operating costs Slow and less efficient process High pressure is needed for compressing biogas and high energy need for water circulation Congestion due to bacterial growth High requirement for water despite regeneration Probable foaming and subsequent H_2S poisoning leading to corrosion Less flexibility toward variations of input gas Additional drying step for removing water from biomethane
Organic scrubbing	Complex process with expensive investment and operation Expensive for small-scale operation High energy needs for solvent regeneration Expensive solvent with difficult handling requirements Loss of operational efficiency due to dilution of glycol into H_2O Difficult regeneration of solvent if H_2S is not reseparated Heating required for complete regeneration
Pressure swing adsorption	Pre-removal of H_2S and H_2O is strongly recommended Trade-of between the purity of CH_4 and quantity of biomethane due to limited selectivity of membranes Not appropriate for high-purity requirements Low CH_4 yield from a single step Multiple modules are required for high purity High electricity requirement for producing each unit of biomethane Moderate-to-high CH_4 losses Possible congestion and contamination of membranes Membranes need to be replaced within 1–5 years
Cryogenic separation	Expensive investment, maintenance, and operation of high-energy requirements Use of various expensive process equipment Contaminant pre-treatment and removal required Technically complex procedure
Membrane separation	Pre-removal of H_2S and H_2O is strongly recommended Trade-of between the purity of CH_4 and quantity of biomethane due to limited selectivity of membranes Not appropriate for high-purity requirements Low CH_4 yield from a single step Multiple modules are required for high purity High electricity requirement for producing each unit of biomethane Moderate-to-high CH_4 losses possible congestion and contamination of membranes Membranes need to be replaced within 1–5 years
Chemical scrubbing	High investment cost Heating is necessary for solvent regeneration High energy requirement Possibility of contaminant accumulation Possible reaction between O_2 and amines, subsequent breakdown and fouling of the amines Risk of brackish coagulation and foaming Corrosion potential Waste chemicals require additional treatment
Biological upgrading	Still in pilot scale, no reliable data on investment, operation, and maintenance cost Limitations and challenges for larger-scale operation is still unknown huge need for large amounts of H2 Risk of free and explosion if the setups are not handled properly Operational parameters, mainly p^H should be regularly monitored and controlled

Table 2.
Challenges of actual biogas upgrading technologies [38].

another issue that lowers the rate of return-on investment by raising the investment and operating costs associated with implementing biogas [22]. Spatial diversity in state-level regulations can lead to inconsistencies and complications for biogas developers in nations with decentralized governance structures [41] . States' differing laws may make compliance more expensive and prevent biogas projects from expanding across regional borders. National regulatory harmonization can encourage more investment and expedite the development process. An additional layer of complexity is introduced by the lack of established definitions and procedures for evaluating biogas potential and project performance [49]. Investors and policymakers find it difficult to assess the true impact of biogas technologies in the absence of a unified framework, which makes cross-state project comparisons difficult. A flourishing biogas industry depends on the creation of efficient support mechanisms such as tax breaks and feed-in tariffs, which can be hampered by this lack of transparency. For biogas to reach its full potential as a renewable energy source, clear and uniform national standards for production, quality, and environmental impact must be established [50].

A number of government agencies and stakeholders are frequently involved in the drawn-out and complicated process of obtaining the required permits and approvals for biogas plants. Permitting delays can discourage potential investors and raise project costs. Similar to this, technical and legal obstacles may make it difficult to connect biogas plants to the gas or electrical grid. Accelerated biogas deployment depends on simplifying permitting processes and enabling grid connection [44]. Inconsistent regulations in various jurisdictions or regions can also impede economies of scale and cause uncertainty. Bureaucratic burdens would be greatly decreased by creating clear, consistent guidelines and harmonizing permitting standards [16]. Public perception can be a problem in addition to regulatory obstacles. Gaining local acceptance requires open communication and active participation in order to address concerns about noise, odor, and possible environmental effects. Positive public perceptions can be promoted by highlighting the advantages of biogas plants, such as lower greenhouse gas emissions, waste management options, and rural economic growth. Lastly, offering project developers financial incentives and technical support can help promote the construction and operation of biogas plants, especially small-scale projects in rural areas. Biogas plants can overcome operational difficulties and initial investment costs with the aid of these support systems [49].

4.5 Social and cultural challenges

One major cultural barrier is the lack of awareness among the general public and substrate holders regarding the advantages of biogas. The economic and environmental benefits of biogas technology are not widely known, which breed resistance and skepticism. Promoting social acceptance and adoption requires increasing awareness through outreach and education initiatives [39]. Another cultural challenge is opposition to waste segregation practices. Organic waste must be separated from other waste types in order for biogas plants to operate efficiently. However, it is challenging to find high-quality feedstock for the production of biogas since many homes and businesses are unwilling to implement waste segregation procedures. A consistent supply of organic waste depends on encouraging waste segregation through rewards and public awareness initiatives [19].

5. Future perspectives in biogas development

In recent years, biogas production and use have drawn a lot of attention as an environmentally friendly and sustainable energy source. Biogas has become a competitive alternative to conventional fossil fuels as a result of the growing emphasis on renewable energy sources and the need to lower greenhouse gas emissions. The creation of more economical and efficient technologies is one of the main outlooks for biogas production in the future. In order to increase the efficiency of biogas production and lower production costs, developments in anaerobic digestion processes, bio-methanation, and organic waste management will be essential. Integrating biogas production systems with infrastructure is another crucial element. Creating biogas plants that are simple to incorporate into municipal, industrial, and agricultural waste management systems is one example of this [54].

Biogas production can be further increased while lessening the environmental impact of waste disposal by making the most of organic waste and byproducts. Regarding application, biogas's potential as a renewable energy source for transportation, heating, and power generation holds the key to its future. Biogas can be transformed into premium bio-methane that can be fed into natural gas pipelines or used as vehicle fuel as long as improvements in biogas purification and upgrading technologies continue [33].

5.1 Technological advancement in biogas production

Technological advancements in biogas production and upgrading aim to substitute natural gas with biomethane. The yield and quality of methane in raw biogas can be improved by optimizing the parameters of the anaerobic digestion system. This includes selecting and treating biogas substrates, conditioning the bacterial community, and utilizing advanced anaerobic digesters, such as high-rate anaerobic biomass velarization using advanced reactors like up-flow anaerobic sludge blanket reactors with biofilm carriers [56]. Biogas application involves upgrading to remove carbon dioxide, hydrogen sulfide, silicons, and other trace compounds from raw biogas, as these pollutants can harm the environment and pose health risks at high concentrations. Additionally, they reduce the energy density of biogas. Therefore, the focus should not only be on producing biogas from organic matter but also on employing suitable and economical upgrading techniques to generate biomethane with over 95% purity for use as a natural fuel substitute. Current upgrading technologies typically remove H_2S, CO_2, and H_2O in separate, costly steps. Recent research has aimed at developing new adsorbents, such as metal oxides, modified mesoporous silicas, activated carbons, and metal-organic frameworks, but these are also expensive. Consequently, efforts are underway to improve the surface properties of natural adsorbents through chemical modifications or treatments for more cost-effective applications. Future upgrading techniques will focus on using cost-effective natural adsorbents and integrating biological methods with conventional upgrading techniques.

Increasing the yield of biomethane from a variety of feedstocks requires improvements in pretreatment techniques. Pretreatment methods such as oxidative, chemical, physical, and thermochemical have demonstrated promise in increasing the production of biogas. Extrusion, for example, has been demonstrated to boost biomethane yield by as much as 190%. Another new field is microbial electrolysis cell (MEC) technology, which provides an efficient and sustainable way to upgrade biogas with little energy consumption. Significant amounts of CO_2 can be absorbed during bio-electro-methanogenesis, which is made possible by MECs, increasing the methane content of biogas [57]. The potential of nanobubble technology to enhance methane production is also being studied. Another new field is microbial electrolysis cell (MEC) technology, which provides an efficient and sustainable way to upgrade biogas with little energy consumption. Significant amounts of CO_2 can be absorbed during bio electro methanogenesis, which is made possible by MECs, increasing the methane content of biogas. The potential of nanobubble technology to enhance methane production is also being studied [58].

5.2 Biogas for circular economy and waste management

Indisputably, recycling resources, energy, and nutrients is a practical strategy to boost process efficiency. Recycling makes it possible to use resources more effectively, which can be enhanced by maximizing raw material and energy consumption. However, the inherent restrictions of thermodynamics prevent limitless recycling of any quantity. According to the second thermodynamic law, there is an inevitable loss in both the quantity and quality of the material during each cycle, making absolute circularity in any transformation process futile [55]. As a result, the concept of a circular economy is more frequently praised than rigorously examined by the scientific community because its execution yields limited and brittle results as it is mostly accomplished globally. When using renewable energy production, the results indicate that it is not acceptable to try to meet all of this demand only through material recycling and renewable energy production. A more thorough examination of cycle efficiency and energy recovery techniques is required. Also, the Circular economy Action Plan does not list using waste materials to generate energy as the main choice.

A crucial component of production lines is lowering the amount of trash generated while also valuing low-quality streams. The circular economy and the sustainability of industrial operations are compatible with this idea. But sometimes, at least not at the ideal rate, it might not be feasible to extract useful goods from waste streams [55]. Because biogas production turns waste into energy and useful byproducts, it is essential to the circular economy. In addition to producing renewable energy, anaerobic digestion of organic waste enhances the manure's fertilizing properties, making it an eco-friendly manure management method [58]. Recovering household waste through composting and biogas production helps to generate renewable energy and waste value. To further close the nutrient loop and lessen dependency on synthetic fertilizers, the digestate nutrient-rich byproduct of the biogas process can be used as biofertilizer. By managing waste holistically, greenhouse gas emissions are decreased, environmental effects are minimized, and sustainable agriculture is encouraged. By encouraging a cooperative relationship between waste management, energy production, and agricultural practices, the application of biogas technology provides a feasible route toward a society that is more resource-efficient and environmentally conscious [7].

5.3 Policy development and marketing strategies

Financial incentives and governmental regulations are essential for encouraging the advancement and application of biogas technologies. To fully utilize biogas technology, cross-sector collaboration and stable, predictable energy policies are required.

National biogas production can be further strengthened by tariffs and subsidies on biogas production facilities [7]. Amid climate change, environmental concerns, and global energy shortages, the biogas sector offers a dual benefit: generating clean energy and managing organic waste. Countries worldwide are focusing on sustainable development and adjusting their energy structures. Over the past 5 years, many have revised laws and policies related to biogas energy growth. Germany has updated its renewable energy law twice during this period, reflecting the EU's biogas policy direction and showcasing advanced biogas technology. The first update in 2014 eliminated subsidies for energy crop-based power and landscape greening waste, while reducing production subsidies for biogas projects, emphasizing the need for medium-sized biogas power plants to compete in the market [12]. In 2017, the renewable energy law was modified to eliminate government pricing mechanisms based on fixed feed-in electricity rates, remove incentives for energy crops and new plant biogas upgrades, and establish a bidding system for the production of renewable energy power [43]. The marketization of power generation from renewable sources has been completely supported by these two measures. Owners of biogas started actively looking for technological ways to increase efficiency.

5.4 Research and development

Ongoing research and development efforts are focused on improving biogas production efficiency and expanding the range of usable feedstocks [2]. Enzyme addition: Studies have shown that pretreatment with enzymes such as amylase and cellulase can increase biogas yield. Microbial communities: Research into the microbial communities involved in anaerobic digestion can help optimize biogas production processes [7]. Future research should explore the integration of biogas with other renewable energy sources, such as solar and wind, to create more resilient and sustainable energy systems. Particularly when solar and wind energy are scarce, biogas can serve as a dependable energy source to supplement other renewable energy sources. In order to optimize energy production and storage based on current weather conditions and energy demand, this integration may entail hybrid systems that combine biogas generators with solar panels and wind turbines [7]. Additionally, research is required to create sophisticated algorithms and control systems that can efficiently manage these integrated systems, guaranteeing smooth energy source switching and optimizing overall efficiency. It is also essential to look into the integrated biogas systems' economic feasibility, including analyses of the costs of the initial investment, ongoing costs, and long-term return on investment. Lastly, future research should look at how widespread biogas integration affects the environment and society, taking into account things such as greenhouse gas emissions, waste reduction, and job creation [18].

6. Conclusions

In summary, the advancement of biogas technology presents a practical solution to the environmental challenges and energy requirements associated with managing organic waste. As discussed in this chapter, biogas serves as a renewable energy source that yields nutrient-rich digestate, lowers greenhouse gas emissions, and contributes to the circular economy. The potential feedstocks for biogas production range from primary organic inputs such as corn silage and liquid co-substrates from agricultural processes (such as manure) to aquatic plants and microalgae, which are favorable for biogas facilities due to their higher lipid and carbohydrate content, enhanced efficiency, and simplified production processes. Various pretreatment methods, including mechanical, microwave irradiation, extrusion, liquid hot water, steam explosion, and chemical treatments (alkaline, acidic, and organosolv), as well as oxidative and biogas upgrading techniques, are utilized to boost methane yield and quality for replacement with natural gas in pipelines. To fully harness their capabilities, biogas systems need to be incorporated into existing energy and waste management frameworks. Although biogas production has progressed significantly, several challenges persist, including policy gaps, financial constraints, and technological barriers. This chapter highlighted the lack of specialized knowledge and technical expertise required for the effective operation and maintenance of biogas reactors, as well as the inadequacy of appropriate pretreatment methods and contamination by heavy metals in anaerobic digestion as technical and infrastructural issues. Economic challenges include the overall investment required for setting up bioreactors and treatment technologies, along with competition from cheaper energy and waste treatment alternatives. Policy and political obstacles consist of a lack of comprehensive national policies, insufficient government commitment, and the poor continuity of previous biogas initiative projects through successive administrations. Additionally, the general public and substrate suppliers are still largely unaware of the benefits of biogas, posing current challenges in the production and utilization sectors. The widespread implementation of biogas plants continues to face obstacles due to their high construction and operational costs, particularly in developing countries. To create a more favorable environment for investment and innovation in biogas technologies, it is essential for both government and private sectors to establish policies and decisions that will enhance biogas generation and scalability. Future forecasts indicate that further research and development will be crucial in expanding the variety of suitable feedstocks and enhancing the efficiency of biogas production; innovations in pretreatment methods, biogas upgrading technologies, and the integration of biogas systems with other renewable energy sources are vital. In conclusion, biogas technology is positioned at the intersection of waste management and energy production.

Acknowledgements

The authors acknowledge the usage of Gemini AI for language polishing of the manuscript.

Conflict of interest

There is no conflict of interest.

Author details

Surafel Argahegn Agdew[1,2*], Melaku Esubalew Cherie[1], Wassihun Guade Alemu[1], Tesfa Nega Gesese[1,2] and Alelign Anmaw Getahun[1,2]

1 Faculty of Chemical and Food Engineering, Bahir Dar Institute of Technology, Bahir Dar University, Bahir Dar, Ethiopia

2 Bahir Dar Energy Centre, Bahir Dar Institute of Technology, Bahir Dar University, Bahir Dar, Ethiopia

*Address all correspondence to: aelmneh@gmail.com; surafel.aregahegn@bdu.edu.et

References

[1] Sravan JS, Tharak A, Mohan SV. Status of biogas production and biogas upgrading: A global scenario. In: Emerging Technologies and Biological Systems for Biogas Upgrading. USA: Elsevier; 2021. pp. 3-26. DOI: 10.1016/B978-0-12-822808-1.00002-7

[2] Papurello D, Santarelli M, Fiorilli S. Physical activation of waste-derived materials for biogas cleaning. Energies. 2018;**11**(9):15-22. DOI: 10.3390/en11092338

[3] Kennedy N, Zhao QB, Ma J, Chen S, Frear C. The selective removal of H2S over CO2 from biogas in a bubble column using pretreated digester effluent. Separation and Purification Technology. 2015;**144**:240-247. DOI: 10.1016/J.SEPPUR.2015.02.046

[4] Id JB, Pielech-przybylska K, Balcerek M, Borowski S, Kregiel D. Integrated bioethanol fermentation/anaerobic digestion for valorization of sugar beet pulp. Energies. 2017;**10**(9):1255. DOI: 10.3390/en10091255

[5] Singh T, Alhazmi A, Mohammad A, Srivastava N, Haque S. Integrated biohydrogen production via lignocellulosic waste: Opportunity, challenges & future prospects. Bioresource Technology. 2021;**338**(July):125511. DOI: 10.1016/j.biortech.2021.125511

[6] Akinbami JK, Ilori MO, Oyebisi TO. Biogas energy use in Nigeria: Current status, future prospects and policy implications. Renewable and Sustainable Energy Reviews. 2001;**5**:97-112

[7] Sher F, Smje N, Sulejmanovi J. Emerging technologies for biogas production: A critical review on recent progress, challenges and future perspectives. Process Safety and Environment Protection. 2024;**188**(June):834-859. DOI: 10.1016/j.psep.2024.05.138

[8] Durán I, Rubiera F, Pevida C. Biogas purification by means of adsorption on pine sawdust-based activated carbon: Impact of water vapor. Chemical Engineering Journal. 2018;**353**(April):197-207. DOI: 10.1016/j.cej.2018.07.100

[9] Ferella F, Puca A, Taglieri G, Rossi L, Gallucci K. Separation of carbon dioxide for biogas upgrading to biomethane. Journal of Cleaner Production. 2017;**164**:1205-1218. DOI: 10.1016/j.jclepro.2017.07.037

[10] Kapoor R, Ghosh P, Kumar M, Vijay VK. Evaluation of biogas upgrading technologies and future perspectives: A review. Environmental Science and Pollution Research. 2019;**26**(12):11631-11661. DOI: 10.1007/s11356-019-04767-1

[11] Rahman MM, Hasan MM, Paatero JV, Lahdelma R. Hybrid application of biogas and solar resources to fulfill household energy needs: A potentially viable option in rural areas of developing countries. Renewable Energy. 2014;**68**:35-45. DOI: 10.1016/j.renene.2014.01.030

[12] Santamouris M, Vasilakopoulou K. Present and future energy consumption of buildings: Challenges and opportunities towards decarbonisation. e-Prime Advances in Electrical Engineering, Electronics and Energy. 2021;**1**(October):100002. DOI: 10.1016/j.prime.2021.100002

[13] Roberts SB, Dallal GE. Energy requirements and aging. Public Health

Nutrition. 2005;**8**(7a):1028-1036. DOI: 10.1079/phn2005794

[14] Chen ZM, Chen GQ. Demand-driven energy requirement of world economy 2007: A multi-region input-output network simulation. Communications in Nonlinear Science and Numerical Simulation. 2013;**18**(7):1757-1774. DOI: 10.1016/j.cnsns.2012.11.004

[15] Maroneze MM, Zepka LQ, Vieira JG, Queiroz MI, Jacob-Lopes E. A tecnologia de remoção de fósforo: Gerenciamento do elemento em resíduos industriais. Revista Ambiente e Agua. 2014;**9**(3):445-458. DOI: 10.4136/1980-993X

[16] Zupančič M, Možic V, Može M, Cimerman F, Golobič I. Current status and review of waste-to-biogas conversion for selected European countries and worldwide. Sustainability. 2022;**14**(3):1-25. DOI: 10.3390/su14031823

[17] International Renewable Eergy Agency. Renewable energy statistics 2024 statistiques d' énergie renouvelable 2024 estadísticas de energía. International Renewable Eergy Agency; 2024

[18] Abanades S et al. A critical review of biogas production and usage with legislations framework across the globe. International journal of Environmental Science and Technology. 2022;**19**(4):3377-3400. DOI: 10.1007/s13762-021-03301-6

[19] Kemausuor F, Adaramola MS, Morken J. A review of commercial biogas systems and lessons for Africa. Energies. 2018;**11**(11):1-21. DOI: 10.3390/en11112984

[20] Parawira W. Biogas technology in sub-Saharan Africa: Status, prospects and constraints. Reviews in Environmental Science and Biotechnology. 2009;**8**(2):187-200. DOI: 10.1007/s11157-009-9148-0

[21] Mutungwazi A, Mukumba P, Makaka G. Biogas digester types installed in South Africa: A review. Renewable and Sustainable Energy Reviews. 2018;**81**:172-180. DOI: 10.1016/j.rser.2017.07.051

[22] Angelidaki I et al. Biogas upgrading and utilization: Current status and perspectives. Biotechnology Advances. 2018;**36**(2):452-466. DOI: 10.1016/j.biotechadv.2018.01.011

[23] Nwokolo N, Mukumba P, Obileke K, Enebe M. Waste to energy: A focus on the impact of substrate type in biogas production. PRO. 2020;**8**(10):1-21. DOI: 10.3390/pr8101224

[24] Karne H, Mahajan U, Ketkar U, Kohade A, Khadilkar P, Mishra A. A review on biogas upgradation systems. Materials Today Proceedings. 2023;**72**:775-786. DOI: 10.1016/J.MATPR.2022.09.015

[25] Ignatowicz K, Piekarski J, Kogut P. Influence of selected substrate dosage on the process of biogas installation start-up in real conditions. Energies. 2021;**14**(18):1-11. DOI: 10.3390/en14185948

[26] Agdew SA, Nega T, Anmaw A. Biogas production: Current status: Challenges: And future perspective:26

[27] Sibiya NT, Tesfagiogis H, Muzenda E. Application of different pre-treatment techniques for enhanced biogas production from lawn grass: A review. In: The Nexus: Energy, Environment and Climate Change. Green Energy and Technology. Cham: Springer; 2018. DOI: 10.1007/978-3-319-63612-2_15

[28] Balat M, Balat H. Biogas as a renewable energy source a review. Energy Sources Part A: Recovery, Utilization, and Environmental Effects. 2009;**31**(14):1280-1293. DOI: 10.1080/15567030802089565

[29] Ignatowicz K, Filipczak G, Dybek B, Wałowski G. Biogas production depending on the substrate used: A review and evaluation study—European examples. Energies. 2023;**16**(2):798. DOI: 10.3390/en16020798

[30] Edem IE, Olugbade QO. Biogas substrates performance ranking based on sustainability potential, biodegradability, and yield characterization: The case of an area in Southwest Nigeria. Nigerian Journal of Technology. 2022;**41**(3):444-453. DOI: 10.4314/njt.v41i3.4

[31] Czekała W, Nowak M, Bojarski W. Characteristics of substrates used for biogas production in terms of water content. Fermentation. 2023;**9**(5):449. DOI: 10.3390/fermentation9050449

[32] Chinnasamy S, Ramakrishnan B, Bhatnagar A, Das KC. Biomass production potential of a wastewater alga chlorella vulgaris ARC 1 under elevated levels of CO_2 and temperature. International Journal of Molecular Sciences. 2009;**10**(2):518-532. DOI: 10.3390/ijms10020518

[33] Prajapati SK, Kaushik P, Malik A, Vijay VK. Phycoremediation coupled production of algal biomass, harvesting and anaerobic digestion: Possibilities and challenges. Biotechnology Advances. 2013;**31**(8):1408-1425. DOI: 10.1016/j.biotechadv.2013.06.005

[34] Mutungwazi A, Ijoma GN, Ogola HJO, Matambo TS. Physico-chemical and metagenomic profile analyses of animal manures routinely used as inocula in anaerobic digestion for biogas production. Microorganisms. 2022;**10**(4). DOI: 10.3390/microorganisms10040671

[35] Tauš P, Kudelas D, Taušová M, Gabániová Ľ. Statistical approach for assessing the suitability of substrates for a biogas plant. Sustainability. 2020;**12**(21):1-14. DOI: 10.3390/su12219044

[36] Struk M, Kushkevych I, Vítězová M. Biogas upgrading methods: Recent advancements and emerging technologies. Reviews in Environmental Science and Biotechnology. 2020;**19**(3):651-671. DOI: 10.1007/s11157-020-09539-9

[37] Garcia NH, Mattioli A, Gil A, Frison N, Battista F, Bolzonella D. Evaluation of the methane potential of different agricultural and food processing substrates for improved biogas production in rural areas. Renewable and Sustainable Energy Reviews. 2019;**112**(August 2018):1-10. DOI: 10.1016/j.rser.2019.05.040

[38] Ahmed SF et al. Biogas upgrading, economy and utilization: A review. Environmental Chemistry Letters. 2021;**19**(6):4137-4164. DOI: 10.1007/s10311-021-01292-x

[39] González R, Peña DC, Gómez X. Anaerobic Co-digestion of wastes: Reviewing current status and approaches for enhancing biogas production. Applied Sciences. 2022;**12**(17):8884. DOI: 10.3390/app12178884

[40] Poeschl M, Ward S, Owende P. Environmental impacts of biogas deployment - part II: Life cycle assessment of multiple production and utilization pathways. Journal of Cleaner Production. 2012;**24**:184-201. DOI: 10.1016/j.jclepro.2011.10.030

[41] Zanatta H, Kanda W, Tonderski K. Biogas Production in Brazil Barriers and Strategies for Overcoming them. Linköping: Linköping University Electronic Press; 2024

[42] Nielsen MB. Identifying challenges and drivers for deployment of centralized biogas plants in Denmark. Sustainability. 2022;**14**(13):8021. DOI: 10.3390/su14138021

[43] Thiruselvi D et al. A critical review on global trends in biogas scenario with its up-gradation techniques for fuel cell and future perspectives. International Journal of Hydrogen Energy. 2021;**46**(31):16734-16750. DOI: 10.1016/j.ijhydene.2020.10.023

[44] Ghosh P, Shah G, Sahota S, Singh L, Vijay VK. Biogas production from waste: Technical overview, progress, and challenges. In: Bioreactors Sustainable Design and Industrial Applications in Mitigation of GHG Emissions. Elsevier; 2020. DOI: 10.1016/B978-0-12-821264-6.00007-3

[45] Grmek M et al. Valencia. 2008;**2008**:16

[46] Guo Q et al. Heavy metals interact with the microbial community and affect biogas production in anaerobic digestion: A review. Journal of Environmental Management. 2019;**240**(February):266-272. DOI: 10.1016/j.jenvman.2019.03.104

[47] Mudhoo A, Kumar S. Effects of heavy metals as stress factors on anaerobic digestion processes and biogas production from biomass. International Journal of Environmental Science and Technology. 2013;**10**(6):1383-1398. DOI: 10.1007/s13762-012-0167-y

[48] Gao M, Wang D, Wang Y, Wang X, Feng Y. Opportunities and challenges for biogas development: A review in 2013-2018. Current Pollution Reports. 2019;**5**(2):25-35. DOI: 10.1007/s40726-019-00106-7

[49] Patinvoh RJ, Taherzadeh MJ. Challenges of biogas implementation in developing countries. Current Opinion in Environmental Science & Health. 2019;**12**:30-37. DOI: 10.1016/j.coesh.2019.09.006

[50] Lu J, Gao X. Biogas: Potential, challenges, and perspectives in a changing China. Biomass and Bioenergy. 2021;**150**(April):106127. DOI: 10.1016/j.biombioe.2021.106127

[51] Tagne RFT, Dong X, Anagho SG, Kaiser S, Ulgiati S. Technologies, challenges and perspectives of biogas production within an agricultural context. The case of China and Africa. Environment, Development and Sustainability. 2021;**23**(10):14799-14826. DOI: 10.1007/s10668-021-01272-9

[52] Yadav P, Yadav S, Singh D, Giri BS. An analysis on generic barriers to bioenergy technologies adoption in context of rural India. Bioresource Technology Reports. 2021;**14**(February):100671. DOI: 10.1016/j.biteb.2021.100671

[53] Masih JC, Sourabh SS, Anand RC. Utilization of organic substrates for biogas production. Current Journal of Applied Science and Technology. 2018;**31**(5):1-12. DOI: 10.9734/cjast/2018/45336

[54] Obileke KC, Nwokolo N, Makaka G, Mukumba P, Onyeaka H. Anaerobic digestion: Technology for biogas production as a source of renewable energy—A review. Energy & Environment. 2021;**32**(2):191-225. DOI: 10.1177/0958305X20923117

[55] Ellacuriaga M, García-Cascallana J, Gómez X. Biogas production from organic wastes: Integrating concepts of circular economy. Fuels. 2021;**2**(2):144-167. DOI: 10.3390/fuels2020009

[56] Getachew A et al. Treatment of water hyacinth juice in an upflow

anaerobic sludge blanket reactor with biofilm carriers. Biomass Conversion and Biorefinery. 2025. DOI: 10.1007/s13399-025-06659-6

[57] Jameel MK et al. Biogas: Production, properties, applications, economic and challenges: A review. Results in Chemistry. 2024;**7**(May):101549. DOI: 10.1016/j.rechem.2024.101549

[58] Swinbourn R, Li C, Wang F. A comprehensive review on biomethane production from biogas separation and its techno-economic assessments. ChemSusChem. 2024;**17**:e202400779. DOI: 10.1002/cssc.202400779

Chapter 2

Comparison of Biogas Production in Six Urban Houses: A Sustainable Approach to Reducing Energy Dependency

Guillermo Yorel Noriega Aquise

Abstract

In the pursuit of a bioclimatic and sustainable home that optimizes thermal comfort and minimizes dependence on external energy, reducing energy consumption and mitigating environmental impact are essential. To achieve this, various techniques and processes are integrated and synchronized within the household. One such process is the production of biogas from the treatment of organic solid waste and biomass generated within the home. This process, developed under adequate thermal conditions, ensures the rapid production of biogas for cooking purposes, with the potential to generate additional energy for the household in an urban setting. The study compares the availability of biogas in urban homes of varying sizes (77.34 m^2 to 179.44 m^2 on 160 m^2 lots) using a vertical and compact biodigester that occupies 1 to 1.5 m^2 and has a volume of 3 to 4 m^3, minimizing spatial impact. The procedures for waste treatment and the amount of biomass generated are detailed, evaluating biogas availability over 1 year and comparing costs with liquefied petroleum gas (LPG). The positive impact of compact biodigesters in urban settings is highlighted, in terms of both environmental sustainability and long-term energy cost reduction. Finally, the scalability of this model is considered adaptable to different urban or rural contexts, enabling its implementation in communities of various sizes and characteristics.

Keywords: single-family housing, sustainable housing, solid waste, family biogas, domestic energy, compact biodigester

1. Introduction

Like many cities in Latin America, Arequipa faces rapid population growth driven by rural migration, leading to disorganized urban expansion, reduced green areas, and overburdened basic services such as water and electricity [1]. With a population of 1,009,132 inhabitants and 301,570 households, solid waste management represents a critical challenge. In 2019, the city generated 262,831.75 tons annually, with significant increases projected by 2034 [2]. Furthermore, only 69.65% of the urban population has access to sewage systems, and wastewater treatment remains insufficient [3].

The inadequate management of solid waste and rising energy costs call for innovative solutions. One such alternative is biogas production in bioclimatic homes, which integrates sustainable technologies to optimize resource use and minimize environmental impact.

Biogas, primarily composed of methane (CH_4) and carbon dioxide (CO_2), is generated through the anaerobic digestion of organic waste. This process, especially efficient in thermally conditioned homes, helps reduce greenhouse gas emissions and dependency on fossil fuels [4]. These bioclimatic homes, designed to leverage local climatic conditions, integrate passive and active technologies to optimize thermal comfort and energy efficiency while managing water and waste sustainably [5, 6].

Bioclimatic design incorporates renewable energies such as solar thermal and photovoltaic systems, reducing costs and emissions [7]. Waste management at the source, composting, and biogas production decrease reliance on landfills and fossil fuels [8]. Furthermore, efficient materials improve air quality and thermal insulation [9].

In urban environments, biogas plays a key role in the circular economy by converting waste into renewable energy and fertilizers, reducing methane emissions, and fostering green jobs [10, 11]. This approach helps lower the carbon footprint of cities while promoting sustainability and energy resilience.

This study evaluates biogas production in urban homes of varying sizes using compact biodigesters. The analysis spans one year, accounting for seasonal variations in production and its impact on waste management and energy consumption.

The main objective is to compare the feasibility of biogas production in urban homes with compact biodigesters as a sustainable strategy to reduce energy dependency, optimize thermal comfort, and mitigate environmental impact.

The specific objectives are:

- Measure biogas production in urban homes of different sizes.
- Assess the efficiency of compact biodigesters in terms of space and functionality.
- Analyze the reduction of energy costs through biogas utilization.
- Study the environmental impact of biogas production.
- Explore the scalability of the model.

The research aims to address the question: Is biogas production in urban homes viable and effective for reducing energy dependency, improving thermal comfort, and mitigating environmental impact? Finally, the hypothesis posited is biogas production through compact biodigesters in urban homes is viable, significantly reducing external energy dependency, improving thermal comfort, and promoting environmental sustainability through efficient waste management.

The proposed system integrates advanced techniques to manage water, energy, and waste in bioclimatic homes. This includes wastewater separation, anaerobic digestion for biogas production, and solar energy harvesting. The benefits encompass reduced energy consumption, efficient waste management, and improved residential comfort.

This chapter specifically analyzes biogas production as a sustainable energy source for cooking in urban homes.

2. Biogas as energy for cooking in family houses

2.1 Perception of biogas in urban houses

Biogas is a renewable gas produced through the anaerobic digestion of organic matter, including food waste, manure, and wastewater. It is primarily composed of methane (50–70%) and carbon dioxide (30–50%), with small concentrations of gases such as hydrogen sulfide and ammonia [12, 13].

The biogas production process involves four stages: hydrolysis, acidogenesis, acetogenesis, and methanogenesis. During the final stage, methanogens convert intermediate compounds into methane and carbon dioxide [14]. Factors such as the type of organic matter, temperature, and digest conditions influence its quality and efficiency [15]. Biogas can be used as fuel for cooking, electricity generation, or heating and can be purified as a substitute for natural gas [16].

In urban settings, perceptions of biogas are changing due to growing interest in sustainability and alternative energy sources. However, its adoption faces challenges related to perceived high costs and the complexity of installation and maintenance [17]. Despite technological advances improving its safety and efficiency, misconceptions persist about its use, often viewed as dirty or hazardous [18].

Biogas plays a critical role in the transition to renewable energy and the circular economy by converting waste into useful energy and reducing dependence on fossil fuels [17]. In urban homes, its use generates clean energy, reduces costs, and lowers the carbon footprint, promoting energy self-sufficiency [19].

However, social and cultural factors influence its adoption. A lack of information and insufficient financial incentives hinder its implementation. To overcome these barriers, educational initiatives, government support, and effective communication about the benefits of biogas are essential [18, 19].

A comprehensive approach can position biogas as a sustainable and viable solution in urban homes, promoting technological adoption and enhancing the environmental and economic well-being of cities.

2.2 Technical aspects of biogas production

Biogas is produced through the anaerobic digestion of organic waste in biodigesters. Common feed stocks include food scraps, animal manure, and agricultural residues. Materials with high fat, sugar, and starch content are the most efficient, while lignocellulosic waste, such as wood and straw, requires more time to break down [20, 21]. Food waste and manure are particularly effective due to their energy content. Ruminant manure is especially suitable, while poultry manure requires careful handling due to its high ammonia content [22]. Agricultural residues, such as corn stover or grass clippings, can be combined with other feedstocks to enhance biogas production [23].

The process begins with waste collection, mixed with water to form a homogeneous slurry before being introduced into the biodigester. Through hydrolysis, acidogenesis, acetogenesis, and methanogenesis, microorganisms decompose organic compounds, generating biogas (CH_4 and CO_2) and digestate, a useful byproduct as fertilizer [20].

Biogas production depends on a balanced carbon-to-nitrogen (C/N) ratio in the feedstock to optimize biodegradability. pH and temperature are crucial factors; a pH range of 6.5 to 8.0 and mesophilic (30–40°C) or thermophilic (50–60°C) temperatures ensure optimal conditions for methanogenic microorganisms [20].

Co-digestion, the combination of different feedstocks, improves reactor stability and increases methane production [24]. Pretreatments such as steam explosion or alkaline hydrolysis break down complex structures, significantly enhancing process efficiency [25].

Phase-separated anaerobic digestion uses distinct reactors for each process stage, optimizing conditions and maximizing methane production [26]. Co-digestion balances the C/N ratio and stabilizes the process, significantly boosting biogas yield [27].

Continuous digestion systems allow for constant biogas production, which is ideal for industrial-scale installations requiring steady gas output [23]. Thermal and chemical pretreatments facilitate the degradation of complex materials such as lignin and cellulose, increasing biogas output, though they require significant initial investment [25].

The efficiency of biogas production depends on feedstock selection, rigorous process condition control, and the implementation of suitable technologies. Strategies such as co-digestion and pretreatments are key to optimizing process performance. The choice of ideal technology depends on available feedstocks, installation size, and production objectives.

2.3 Technology and equipment required for biogas production

Anaerobic digestion for biogas production requires specific technologies and equipment that ensure efficiency and high productivity.

The anaerobic digester is the central element of the system, responsible for the decomposition of organic matter in the absence of oxygen. Its design varies depending on the operating temperature, being classified as mesophilic (30–38°C) or thermophilic (50–60°C). Thermophilic digesters are more efficient in the destruction of pathogens and greater biogas production, although more expensive. Depending on the solid content, digesters can be for dry processes (high solids content) or wet processes (low solid content). Dry digesters are suitable for more solid waste, while wet digesters are used for liquid or slurry waste. Digesters can be continuous flow for constant feeding and continuous production, or batch for specific operations [28].

Pretreatment equipment improves the efficiency of anaerobic digestion. This includes crushers and pulverizes that reduce particle size, increasing the surface area available for microbial action. Hydrolyzers release sugars from lignocellulosic waste, facilitating its fermentation [25].

Biogas purification and upgrading systems remove impurities to improve its quality and expand its applicability. Membrane separation allows methane to be extracted, obtaining biomethane suitable for vehicle fuels or for injection into natural gas networks. Scrubbing systems remove contaminants such as hydrogen sulfide and water, reducing the risk of corrosion [29].

In addition to the main components, complementary equipment is required, such as biogas storage systems that allow gas management before use or marketing. Internal combustion engines or microturbines are used to generate energy, which converts biogas into electricity and heat [28].

Digestate treatment systems process the resulting solid or liquid waste, allowing it to be used as agricultural fertilizer and closing the nutrient recycling cycle [29].

The use of appropriate technologies and specific equipment ensures an efficient and sustainable process for biogas production. From digester design to biogas purification and digestate handling, each component plays a crucial role in maximizing productivity and minimizing environmental impact.

2.4 Biogas production efficiency and yield

Biogas production efficiency and yield are essential to maximize energy generation and ensure the economic viability of the process. Key factors influencing these parameters include the following:

- Temperature and pH: Mesophilic conditions (30–38°C) are widely used, while thermophilic conditions (50–60°C) accelerate digestion and eliminate pathogens, although they require more rigorous control. The optimum pH for methanogenic bacteria activity is between 7.0 and 7.5.

- Hydraulic retention time (HRT): An adequate HRT ensures complete digestion of the substrates. However, excessively long retention times can negatively affect the economic efficiency of the system [30].

- Biogas yield is optimized by measuring the amount of methane produced per unit of degraded organic matter. Strategies to maximize this include:

- Substrate pretreatment techniques such as grinding, thermal hydrolysis, and steam explosion are considered to increase the accessibility of substrates for microorganisms, which improves the biogas production rate [31].

In co-digestion, the combination of different types of organic waste balances the nutrients in the digester, increasing both methane production and process stability. The use of advanced technologies, such as membrane separation to purify biogas and anaerobic digestion in separate phases, improves process control and maximizes the conversion of organic matter to methane [28].

Optimizing efficiency and performance in biogas production requires precise control of process parameters, effective substrate pretreatment, and the incorporation of advanced technologies. These strategies increase methane production, improve biogas quality and enhance system viability.

Incorporating biogas into bioclimatic and sustainable housing design improves energy efficiency, reduces the carbon footprint, and encourages a self-sufficient and environmentally friendly housing model. Bioclimatic design focuses on optimizing the orientation and location of the home to maximize the use of local climate conditions. Locating the biogas system near the kitchen or composting areas simplifies the management of organic waste and improves system efficiency [32].

The concept of closed energy and nutrient cycles allows biogas, produced from the anaerobic digestion of organic waste (such as kitchen scraps and wastewater), to be used for cooking, heating water, or generating electricity. The remaining digestate acts as a fertilizer, closing the nutrient cycle and reducing the use of chemicals [33].

Biogas complements sources such as solar and wind energy, ensuring a constant energy supply. For example, photovoltaic panels can cover daytime needs, while biogas is stored for nighttime use or cloudy days, maximizing energy self-sufficiency [34].

The use of biogas decreases dependence on fossil fuels and significantly reduces greenhouse gas emissions. Furthermore, as a local and small-scale system, biogas minimizes energy losses associated with distribution, improving overall efficiency [33].

Bioclimatic architecture adapts to various climatic conditions, and biogas plays specific roles depending on the environment: in cold climates, as a heating source; in hot climates, for electricity generation and cooking [32, 33].

Biogas and bioclimatic architecture form an effective combination that offers a versatile and sustainable solution, optimizing the use of resources based on the specific needs of each climatic environment. This integration is essential for the development of more efficient and environmentally responsible housing.

2.5 Isolation and safety of biogas production

Isolation and safety in biogas production in urban homes requires exploring the best practices, technologies, and regulations that guarantee the safety of the inhabitants and the environment.

The location of the biodigester must comply with regulations that ensure its installation in a well-ventilated area, preferably outside the home, to minimize the risk of gas accumulation. In many countries, there are regulations that specify the minimum distance that must be maintained between the biodigester and habitable areas [35]. It is advisable to establish physical barriers around the biodigester to protect it from external impacts or interference, using fences, concrete walls, or metal structures [36].

The construction materials of the biodigester must be resistant to internal gas pressure and corrosion caused by organic waste. Materials such as stainless steel, fiberglass-reinforced plastics, or coated concrete are common options [35].

Waterproofing and sealing are essential to prevent leaks; seals should be checked periodically to ensure the absence of gas leaks [37, 38].

Leak prevention, which includes installing gas sensors near the biodigester, is an effective measure to detect possible methane leaks. These sensors should be connected to alarms to alert residents in the event of a leak [39]. Regular preventive maintenance, including checking valves, joints and connections, is crucial to prevent leaks [37].

Handling flammable gas requires an adequate ventilation system, which allows any accidentally escaping gas to be quickly dispersed [36]. It is essential to keep the biodigester and biogas storage tanks away from ignition sources such as stoves, heaters, or electrical connections [39].

Evacuation plans should include training residents in procedures for handling gas leaks. It is important to have a clearly defined emergency plan [35]. It is important to install fire extinguishers designed for gas fires near the biodigester and biogas storage areas [39].

The installation of biogas systems must comply with local and international safety regulations, including certifications and periodic inspections by competent authorities. In some cases, certifications or periodic inspections by competent authorities are required.

Biogas control systems must be modern and up-to-date, incorporating automated monitoring to monitor pressure, temperature, and gas levels in real time. The use of automated monitoring systems that monitor pressure, temperature, and gas levels in real time can prevent dangerous situations. Such systems can include automatic shut-off mechanisms for emergency situations [40].

Insulation and safety in biogas production require careful design, the use of resistant materials, constant monitoring, and compliance with regulations. These measures ensure a safe and efficient environment, minimizing the risks associated with biogas management in urban homes.

2.6 Environmental impact and sustainability of biogas production

Biogas production in urban homes contributes significantly to environmental sustainability by offering a renewable energy alternative and promoting efficient waste management.

Converting household waste into biogas reduces the amount of municipal solid waste reaching landfills, decreasing greenhouse gas (GHG) emissions associated with its decomposition [41, 42]. Safe processing of organic waste prevents the leaching of harmful compounds that could contaminate soil and groundwater [43].

Biogas replaces fossil fuels in cooking and heating, reducing CO_2 and other GHG emissions, thereby decreasing dependence on fossil fuels [44]. Capturing and using methane generated by organic waste prevents its release into the atmosphere, which is especially beneficial given that methane has a higher GHG impact than CO_2 [45].

Unlike fossil fuels, biogas combustion generates less particulate matter and nitrogen oxides (NOx), improving air quality and reducing respiratory health risks in urban areas [46].

Digestate, a byproduct of biogas production, acts as a natural fertilizer in urban gardens and orchards, closing the nutrient cycle and reducing the need for synthetic fertilizers that can pollute aquatic ecosystems [47].

Biogas production in urban homes promotes sustainable waste management, reduces GHG emissions, improves air quality, and contributes to nutrient recycling using digestate as a fertilizer, establishing a circular model that benefits both the environment and urban communities.

3. Methodology

3.1 Comparative methodological considerations

The comparative method is an essential tool in scientific research, supported by authors such as Lasswell [48], Almond [49], Nohlen [50], and Smelser [51]. This method facilitates the analysis of similarities and differences between cases or variables, promoting empirical generalization and hypothesis validation [52, 53].

The advantages of comparative methods include understanding unknown phenomena and extrapolating knowledge from well-studied phenomena to less understood contexts. It aids in explaining and interpreting phenomena, fostering the development of new knowledge through the comparison of similar cases. The method organizes information systematically, highlighting peculiarities between compared phenomena and enabling the formulation of theories and concepts based on empirical generalization through the analysis of common patterns.

Harold Lasswell [48] highlighted the use of the comparative method to analyze power dynamics and political structures, identifying recurring patterns across different contexts. Gabriel Almond [49] emphasized its application in studying political development, comparing variables across cultures and geographies to understand key factors in the evolution of political systems. Dieter Nohlen [50] underlined the method's relevance for empirically generalizing and verifying hypotheses in social sciences through systematic comparisons based on solid evidence. Neil Smelser [51] advocated for a detailed and rigorous approach to the comparative analysis of economic activities, aimed at identifying patterns and developing applicable theories.

In this research, the comparative method evaluates various techniques and procedures related to the construction and management of sustainable housing. This approach enables comparisons of variables such as energy consumption, wastewater treatment efficiency, biogas production, and electricity generation.

Comparative analysis is crucial for identifying more effective and sustainable practices, contributing to the development of housing models that optimize resource use and minimize environmental impact. By organizing information systematically and highlighting differences and similarities, the comparative method enables robust conclusions based on empirical evidence [50, 54, 55].

3.2 Sequential procedure

The sequential procedure to develop and evaluate biogas production using compact biodigesters in urban dwellings follows a structured approach. The procedure begins by identifying urban dwellings with representative characteristics of different sizes to design the compact biodigester, optimized for biogas production [56]. Then, it continues as follows.

Definition of the comparative method, a comparative approach is applied to analyze the techniques in observation units (OU), evaluating the specific characteristics of each urban dwelling. This method allows the establishment of the basis for a structured and exhaustive analysis.

Identification of main variables, energy needs, the characteristics of organic waste and the design parameters of the biodigester are determined. These variables include the average energy consumption, the type of waste available, and the required capacity of the system [57].

Assessment of biogas needs, based on the identified variables, the specific biogas needs are calculated for each OU. This step includes the analysis of energy demand and the volume of waste generated by each household. Raw material preparation includes processes such as selection, crushing, adjustment of the carbon/nitrogen ratio (20:1 to 30:1), and pre-humidification to maximize nutrient availability and improve methane production.

Biodigester design and sizing of compact biodigesters are designed considering the identified needs. Technical sizing includes calculations for production capacity, optimal operating temperature, and hydraulic retention time (HRT). In addition, resistant materials such as high density polyethylene (HDPE), polyvinyl chloride (PVC), and fiberglass are selected, guaranteeing durability and efficiency [57, 58].

Biogas production, production processes are analyzed under real operating conditions. This includes biogas production tests and design adjustments to maximize efficiency in different scenarios. Biochemical processes such as hydrolysis, acidogenesis, acetogenesis, and methanogenesis are optimized to improve the performance of the biodigester [58].

Biogas production and need relationship, the balance between estimated production and the energy needs of each home is evaluated. This analysis allows identifying possible adjustments in the design and operations to ensure a constant supply [59].

Environmental impact, the environmental impact of the system is analyzed, including the reduction of urban solid waste and the mitigation of greenhouse gas emissions. The use of digestate as an organic fertilizer is also evaluated, promoting the circular economy [59].

Energy cost reduction and the energy costs before and after the installation of the biodigester are compared. The analysis includes the reduction of liquefied petroleum gas (LPG) consumption and its impact on the family budget, highlighting the economic benefits of the system. Previous studies have shown a significant reduction in energy costs for urban homes [60].

Scalability and technology adoption explores the possibilities of implementing the system on a larger scale in different urban contexts. This step includes an analysis of economic, social, and technical feasibility to encourage mass adoption of technology [61].

The sequential procedure covers the design, implementation, and evaluation of compact biodigesters in urban homes. This approach facilitates the analysis of their technical, economic, and social feasibility, offering a sustainable model for biogas production in urban environments (**Figure 1**).

3.3 Units of study

The study units are single-family homes located in Arequipa, Peru, selected according to the National Building Regulations (RNE). The homes follow the type 3 model, with 160 m^2 plots and built areas reflecting the characteristics of the urban environment.

These homes represent diverse residential configurations typical of the city, all with access to basic water and sanitation systems.

The built areas vary depending on the number of residents, demonstrating a progression in land use as population density increases. The homes are situated in consolidated or developing urban areas, reflecting the actual conditions of the population (**Table 1**).

This diagram illustrates the relationships between built areas, free areas, and the number of residents, facilitating the analysis of the impact of design on space utilization and the functionality of the residential environment (**Figure 2**).

3.4 Applications of the methodology

3.4.1 Biogas production in urban homes of different sizes

According to the 2020 National Household Survey (ENAHO), prepared by the National Institute of Statistics and Informatics (INEI) of Peru, LPG is the primary fuel used in Peruvian households for cooking. Approximately 80% of households in Peru use this hydrocarbon [62]. The regions of Arequipa (93.7%), Tacna (92%), and Tumbes (90.1%) report the highest levels of use [63].

At the national level, Osinergmin, the Supervisory Agency for Investment in Energy and Mining of the Peruvian Government, published a report on the Results, Consumption, and Uses of Liquid Hydrocarbons and LPG. It states that the average monthly consumption of LPG is 1.05, with 10-kg cylinders per household, slightly higher in urban areas, at 1.06 cylinders [63]. The average number of household members is 3.2 [62]. This equates to an average per capita LPG consumption of 3.31 kg per month.

The equivalent biogas consumption is determined based on the following calculations:

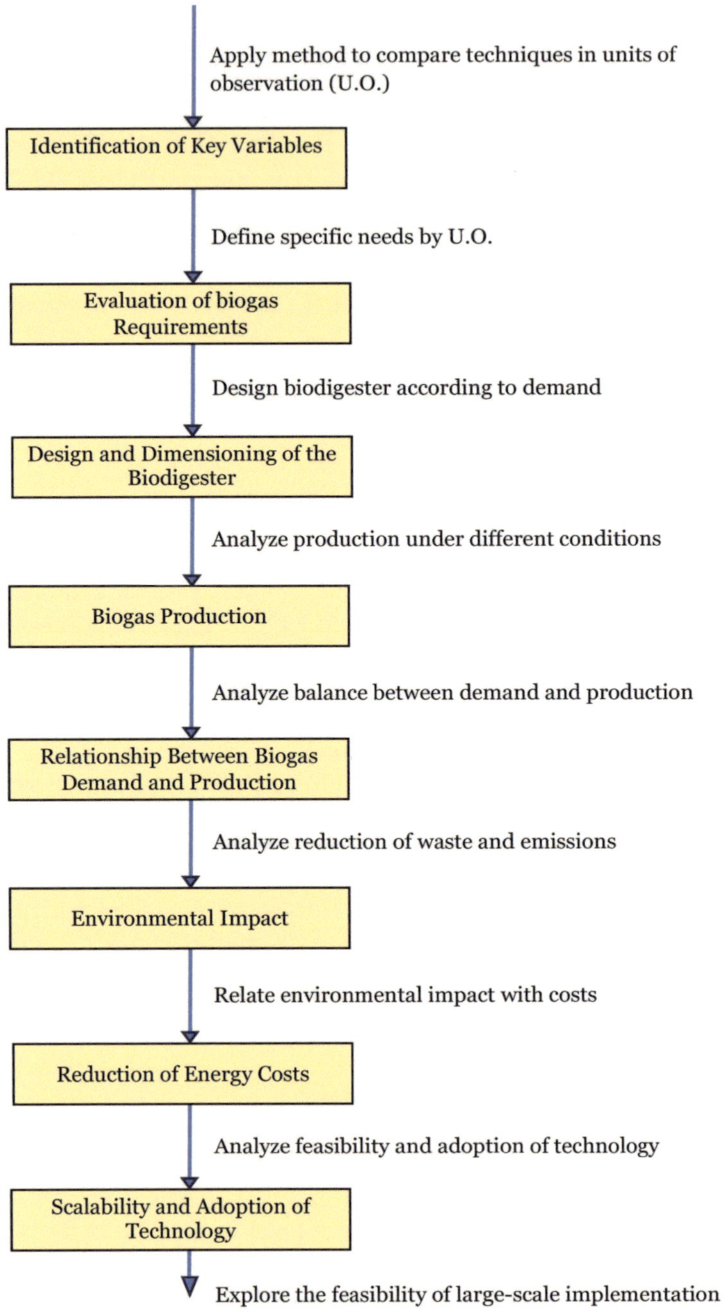

Figure 1.
Block flow of the methodological process.

The average LPG consumption is 3.31 kg per person per month.

1 kg of LPG is equivalent to 1.975 liters of LPG.

1 liter of LPG equals 23,091 british thermal unit (BTU).

Using these conversions, it is calculated that 1 kg of LPG corresponds to 45,604.73 BTU/kg.

The average energy consumption per person is calculated as:

Home ID	Total land area (m^2)	Length (m)	Width (m)	Residents	Total built area in two levels (m^2)	Free area (m^2)
VUF_2-01	160	8.48	4.56	4	77.34	121.33
VUF_2-02	160	8.48	6.38	7	108.20	105.90
VUF_2-03	160	9.40	6.38	7	119.94	100.03
VUF_2-04	160	8.93	7.68	8	137.16	91.42
VUF_2-05	160	9.18	8.48	8	155.69	82.16
VUF_2-06	160	10.58	8.48	9	179.44	70.28

Table 1.
Characteristics of study homes.

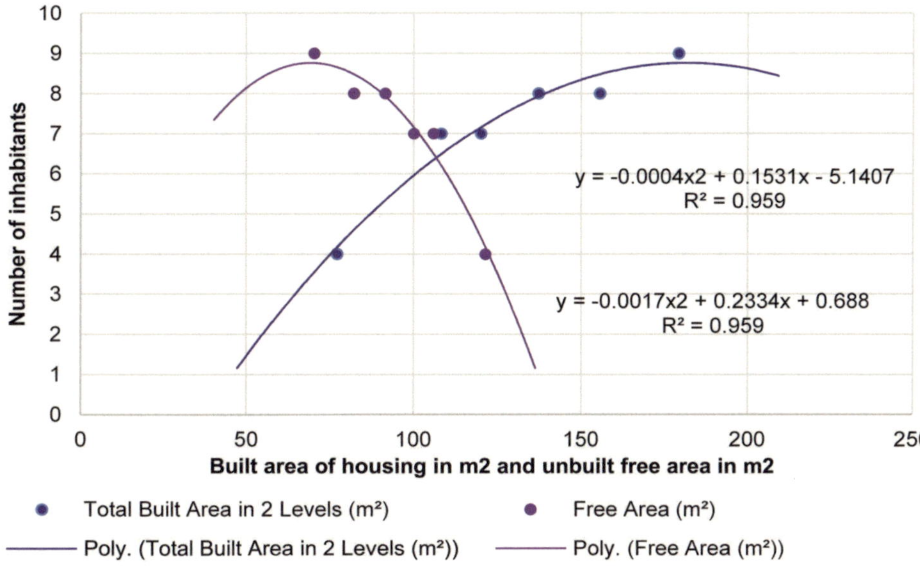

Figure 2.
Built and unbuilt residential areas by number of residents.

3.31 kg/month × 45,604.73 BTU/kg = 150,951.64 BTU/month.

For biogas, 1 cubic meter contains between 18,000 and 22,000 BTU/m^3; for this calculation, an average value of 20,000 BTU/m^3 is used. The energy consumption is then derived from the ratio of BTU from LPG to BTU from biogas:

150,951.64 BTU/20,000 BTU/m^3 = 7.22 m^3 of biogas per person per month.

In this study, it is estimated that a person consumes 5 kg of LPG, where 3.31 kg per person/month is allocated for cooking, and the remaining 1.69 kg per person/month is used for other purposes.

All biogas design calculations are therefore based on 5 kg of LPG per person/month under the conditions of Arequipa, Peru. With 1 kg of LPG equivalent to 45,604.73 BTU, 5 kg LPG corresponds to 228,023.63 BTU/person/month.

The biogas consumption in cubic meters is established as follows:

228,023.63 BTU/20,000 BTU/m^3 = 11.40 $m^3$of biogas per person per month which translates to 0.38 m^3/person/day.

Items	Description	Quantity	Unit
1.0.	Data		
1.1.	Daily biogas consumption for a family of five members	1.90	m^3/day
1.2.	Biogas production: for 0.40 m^3, 1.35 kg of organic waste is required		
1.3.	Daily biomass load:	5.07	kg/day
1.4.	Hydraulic retention time (HRT):	25.00	Days
2.0.	Digester volume		
2.1.	Total biomass load:	126.72	kg
2.2.	kg of organic waste = 1 liter (0.001 m^3) of volume		
2.3.	Total digester volume: (100%)	0.1690	m^3
2.4.	Digester volume = 126.72 kg × 0.001 m^3/kg = 0.1267 m^3 (75%)	0.1267	m^3
2.5.	Space for gas volume (25%)	0.0422	m^3

Table 2.
Key data for compact biodigester sizing.

According to the biogas guide from the United Nations Energy Resources Development Series, New York, 1980 [64], the average gas consumption for cooking meals per person per day is between 0.28 and 0.42 m^3. The value used in this study of 0.38 m^3/person/day falls within this range.

For a family of five, the daily biogas requirement is 1.90 m^3, equivalent to 57.01 m^3 per month (**Table 2**).

The FAO Biogas Manual, 2011 [65], establishes a biogas production rate of 0.40 m^3 of biogas per kg of waste. Based on this rate, a digester for a family of five should produce 1.90 m^3 of biogas/day, requiring 4.75 kg of organic waste per day.

The biogas production rate of 0.40 m^3 per kg of organic waste is achieved under optimal conditions at 35°C. However, the average temperature in Arequipa is 22.69°C. Consequently, the biogas production efficiency is calculated as:

$$\text{Efficiency} = \left(1 - (\text{Optimal Temp} - \text{Internal Temp}) / \text{Optimal Temp}\right)$$

This results in an efficiency level of 0.6483. To achieve maximum efficiency, the required kilograms of waste are adjusted to 0.3517 compared to 1. This value corresponds to 0.35 kg of organic waste. Under Arequipa's average temperature conditions (22.6°C), 1.35 kg of organic waste is needed to produce 0.40 m^3 of biogas. Thus, a family of five must produce 1.90 m^3 daily, requiring 2.57 kg of organic waste.

Annually, a family of five requires 1850.16 kg/year of organic waste. Monthly, the requirement is 154.18 kg/month, and the daily requirement is 5.07 kg/day.

With this base data, a compact vertical biodigester can be sized for cooking food for a family of five.

With this base data, a compact vertical biodigester is sized to cover the cooking needs of a family of five (**Table 3**).

3.4.2 Detailed description of the compact biodigester

The compact biodigester is designed to optimize biogas production in small spaces, integrating mechanical rotating blades that improve the digestion process and guarantee high efficiency.

Items	Description	Quantity	Unit
1.0.	Pre-digestion chamber		
1.1.	Height of the pre-digestion chamber	44.22	cm
1.2.	Length of the pre-digestion chamber	1.83	m
1.3.	Diameter of the digester	44.22	cm
1.4.	Cylindrical volume: $V = \pi \times r^2 \times h$	0.28	m^3
1.5.	Volume of the pre-digestion chamber	0.25	m^3
1.6.	Free volume	0.03	m^3
1.7.	Radius of the pre-digestion chamber	0.22	m
2.0.	Digestion chamber (one unit)		
2.1.	Height of the digester	1.10	m
2.2.	Cylindrical volume: $V = \pi \times r^2 \times h$	0.169	m^3
2.3.	Radius of the digested chamber	22.11	cm
2.4.	Diameter of the digestion chamber	44.22	cm
3.0.	Total digestion chamber (four units)		
3.1.	Height of the digester	1.10	m
3.2.	Cylindrical volume: $V = \pi \times r^2 \times h$ (one chamber)	0.169	m^3
3.3.	Cylindrical volume: $V = \pi \times r^2 \times h$ (four chambers)	0.680	m^3
3.4.	Radius of the digested chamber	22.11	cm
3.5.	Diameter of the digestion chamber	44.22	cm
4.0.	Effluent reception chamber (two units)		
4.1.	Height of the effluent reception chamber	32.40	cm
4.2.	Radius of the effluent reception chamber	0.16	m
4.3.	Length of the effluent reception chamber	1.83	m
4.4.	Volume of the effluent reception chamber	0.30	m^3

Table 3.
Dimensions of the vertical compact biodigester.

In the design of the compact biodigester, the typical spatial restrictions of urban environments were successfully adapted, having to occupy reduced spaces, which is what is considered in the recommendations of Bond and Templeton (2011) [66].

A detailed description of each component of the system is presented below.

Pre-digestion chamber: Its dimensions are 44.22 cm high, 1.83 m long and 44.22 cm in diameter, with a total volume of 0.28 m^3, of which 0.25 m^3 (90%) corresponds to the pre-digestion volume and 0.03 m^3 (10%) to the free volume. This chamber receives the biomass and carries out an initial anaerobic decomposition process. At this stage, the acidity is reduced and the preparation of the biomass for the following stages is improved, and the risk of obstruction is reduced by avoiding the formation of unwanted solids (**Figure 3**).

Digestion chamber (digesters), The design includes four digesters, each 1.10 m high, 44.22 cm in diameter and with a volume of 0.169 m^3, adding up to a total volume of 0.68 m^3. This chamber is the core of the system, where the main anaerobic

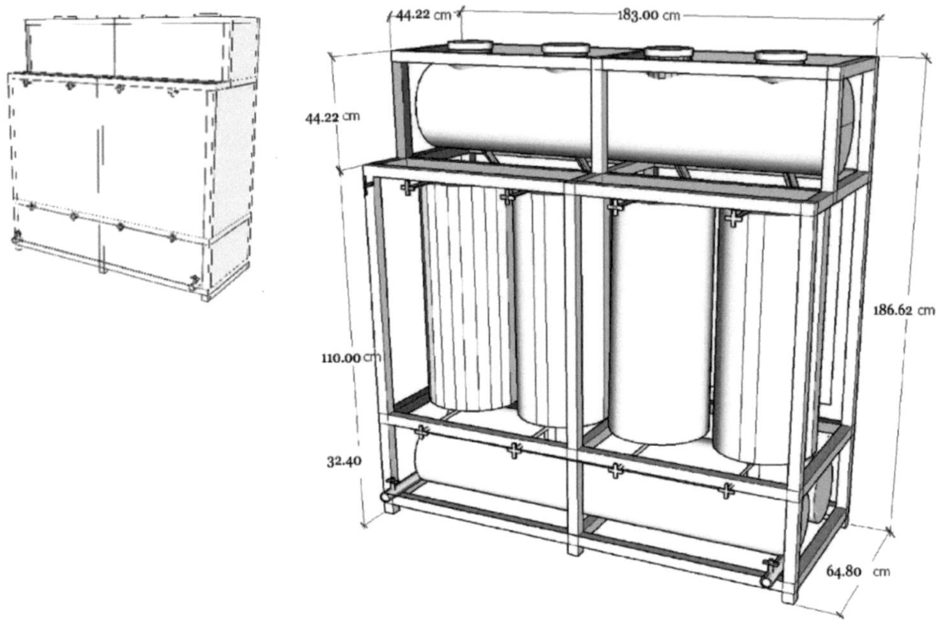

Figure 3.
Compact vertical biodigester design.

digestion occurs. In this stage, bacteria efficiently break down the biomass to produce methane. Each digester houses 0.1267 m^3 of biomass, reserving 25% of its volume for gas accumulation. The installed mechanical rotating blades ensure a homogeneous mixture of the biomass and prevent sedimentation. The benefits allow for increased contact between the biomass and the bacteria. These blades also promote greater gas production and prevent blockages that could interrupt the flow of gases. The effluent reception chamber (digestate), its dimensioning considers two chambers, height 32.40 cm, radius 0.16 m, length 1.83 m, and total volume 0.30 m^3. This chamber collects and stores the final digestate, a nutrient-rich byproduct that can be used as organic fertilizer. This allows a controlled flow from the digesters, maintaining the necessary pressure for biogas production. It allows the periodic collection and disposal of the digestate without interrupting the digestion process.

The biodigester cover provides structural protection to the system, maintains stable temperature conditions, prevents the entry of contaminants, and offers a compact and esthetic appearance suitable for urban environments.

Operation, production, and functioning data. The daily biogas consumption is 1.90 m^3, sufficient to cover the needs of a family of five people. Biogas production is 0.40 m^3 for every 1.35 kg of organic waste. The daily biomass load required is 5.07 kg of organic waste. The hydraulic retention time (HRT) is 25 days per digester.

Daily biomass loading initially passes through the pre-digestion chamber, where its acidity is reduced, and it is prepared for the main process. Anaerobic digestion takes place in the digestion chambers, where the biomass is broken down and biogas is released, a process that is optimized by the rotating blades. The gas accumulates at the top of the digesters and is extracted for domestic use.

The collection of digestate is the effluent that is directed to the reception chambers, facilitating its disposal and use as fertilizer.

The compact biodigester design integrates efficiency, functionality, and space optimization, offering a sustainable solution for biogas production in urban areas. This design guarantees maximum energy production from organic waste, promoting energy savings and reducing waste generation.

4. Results

4.1 Need for biogas per built surface

The need for biogas responds to the growing demand for sustainable, economic, and accessible energy sources in a global context of energy transition.

The energy crisis and diversification of sources, since the dependence on fossil fuels generates long-term sustainability challenges, such as high costs and environmental problems. In this framework, biogas is presented as a renewable source that takes advantage of organic waste and contributes to energy security.

There is an urgent need to carry out organic waste management; the inadequate management of organic waste causes environmental problems, such as greenhouse gas emissions and soil and water contamination. Biogas allows these wastes to be valorized, converting them into useful energy and reducing their environmental impact.

It allows achieving a socioeconomic impact, in communities with limited access to conventional energies, biogas provides an affordable alternative to meet basic energy needs, promoting self-sufficiency and improving the quality of life.

It allows us to contribute to climate change mitigation, and the production and use of biogas reduces methane emissions, a greenhouse gas with a significant impact, which would otherwise be released by the anaerobic decomposition of organic waste.

In this perspective, according to our study units, the need for biogas is shown in **Table 4**.

Table 4 shows that:

As for the land area, all the houses have an area of 160 m^2, with two-story buildings that vary between 77.34 m^2 and 179.44 m^2. This shows different designs or building construction sizes. It shows that the greater the constructed area, the less free area available on the land.

The relationship between the number of inhabitants and the need for biogas is proportional: Houses with more inhabitants require greater volumes of biogas. The need for biogas is directly related to the number of inhabitants, so for four people 25 m^3/month is required, for seven people 43.75 m^3/month is required, for eight people 50 m^3/month is required, and for nine people 56.25 m^3/month is required.

The table suggests that the design of homes should balance the built area, the free area and the need for resources such as biogas, depending on the size of the family, and the free area allows for obtaining organic matter such as grass for biogas production.

Attention should be paid to the importance of planning homes not only in terms of building space and free area, but also considering the capacity to generate and provide sustainable resources for biogas production. For larger families, it would be necessary to consider more efficient biodigesters or those with a greater capacity to meet the needs (**Figure 4**).

The curve shows an increasing trend that flattens out as the built-up area increases. This indicates that the biogas requirement grows rapidly for small areas, but the increase slows down from a certain building size.

Home	Total land area (m^2)	Constructed area (m^2)	Free area (m^2)	Inhabitants	Biogas requirement (m^3/month)
VUF_2-01	160	77.34	121.33	4	25
VUF_2-02	160	108.2	105.9	7	43.75
VUF_2-03	160	119.94	100.03	7	43.75
VUF_2-04	160	137.16	91.42	8	50
VUF_2-05	160	155.69	82.155	8	50
VUF_2-06	160	179.44	70.28	9	56.25

Table 4.
Relationship between constructed area, inhabitants, and biogas requirement.

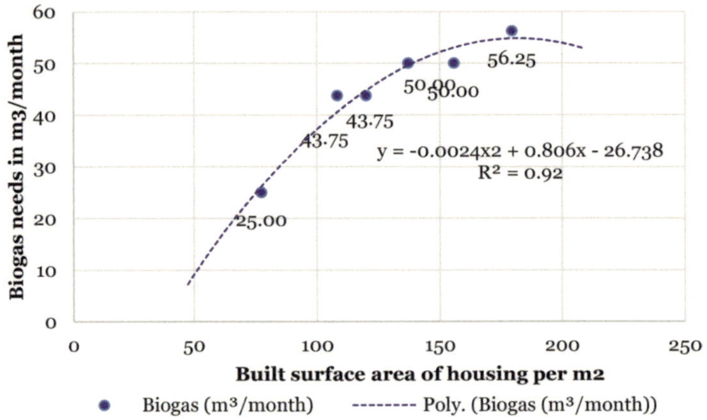

Figure 4.
Relationship of biogas requirement per built-up area.

The analysis of the cubic curve is represented by the polynomial equation:

$$y = -0.0024x2 + 0.806x - 26.738$$

Where x is the built-up area (m^2) and y is the biogas requirement (m^3/month). The coefficient of determination R2 = 0.92 indicates an excellent fit.

The trend of the curve shows that initially, the curve shows an increase in biogas requirement as the built-up area increases. Having reached a maximum around $x \approx 160\ m^2$, the curve begins to decrease, showing that the biogas requirement decreases for larger areas. As the surface area of the dwelling increases, the need for biogas increases due to the expected higher energy consumption. However, after a certain size, energy efficiency can stabilize or existing systems cover the demand, decreasing the need for biogas.

Dwellings with a larger built-up area (such as VUF2-06 with 179.44 m^2) reach the highest need for biogas (56.25 m^3/month). The peak point of biogas need is reached with a built-up area of approximately 167.92 m^2, where the demand is 40.93 m^3/ month.

For built-up areas of approximately 70 m^2 to 120 m^2, the need for biogas grows noticeably. Between 120 m^2 and 180 m^2, the curve begins to stabilize, indicating that the increase in biogas need becomes less pronounced.

The curve indicates that although there is a direct relationship between built-up areas and biogas needs, this is not strictly proportional. Factors such as the efficiency of the design of larger homes influence this trend.

The graph suggests that when designing larger or more populated homes, it is important to consider that biogas demand may not increase indefinitely but may reach a practical limit. This could be useful for optimizing the capacity of biodigesters based on the actual needs of each home (**Figure 5**).

The relationship between the variables seems to remain constant throughout the analysis range, probably because almost all homes require a regular need for biogas. The number of inhabitants grows from 4 to 9, which proportionally increases the need for biogas. The number of inhabitants grows in a staggered manner (4, 7, 7, 8, 8, 9); this variation is directly associated with the demand for biogas, since more inhabitants require more energy generation.

The need for biogas increases consistently with the number of inhabitants; homes with four inhabitants require less biogas and homes with nine inhabitants show the highest consumption. The need for biogas increases as the total built area increases. This is consistent with the logic that larger homes tend to have higher energy demand.

In segmental behavior, from 77.34 m^2 to 108.2 m^2, the need for biogas increases significantly from 25 m^3/month to 43.75 m^3/month. From 108.2 m^2, the growth in the need for biogas became more stable, increasing from 43.75 m^3/month to 56.25 m^3/month in 179.44 m^2.

The need for biogas is directly influenced by the number of inhabitants and the built surface area. The graph highlights the importance of planning biogas systems based on the density of occupation of the homes, since the energy demand does not depend so much on the physical space as on the use of the home.

Table 5 shows the relationship between ambient temperatures and biogas production in a system with one or two digesters throughout the months of the year.

The average temperature allows biogas production to fluctuate with the average monthly temperature: It reaches maximum values in warm months (16.5°C) and

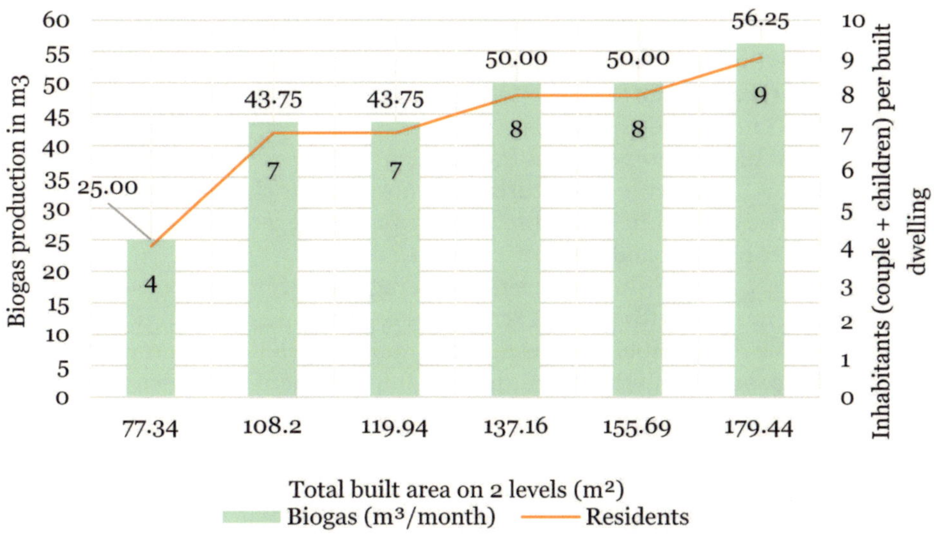

Figure 5.
Relationship between land area, biogas needs, and inhabitants.

Month	Maximum temperature °C	Minimum temperature °C	Average temperature °C	Biogas in m^3 at average temperature °C (one digester)	Biogas in m^3 at average temperature °C (two digesters)
January	22.1	10.9	16.50	41.45	82.90
February	22.1	11.2	16.65	41.83	83.65
March	22.3	10.6	16.45	41.32	82.65
April	22.8	9.8	16.30	40.95	81.90
May	22.8	8.2	15.50	38.94	77.88
June	22.8	7.2	15.00	37.68	75.36
July	22.6	6.9	14.75	37.05	74.11
August	22.9	7.8	15.35	38.56	77.12
September	23.1	8.4	15.75	39.57	79.13
October	23.2	9.1	16.15	40.57	81.14
November	23.0	9.4	16.20	40.70	81.39
December	22.6	10.2	16.40	41.20	82.40

Table 5.
Biogas production as a function of temperature in Arequipa.

minimum values in cold months (14.75°C). In warmer months (such as January and December, with average temperatures of ~16.5°C), biogas production reaches maximum values. In colder months (such as July, with 14.75°C), production decreases. Biogas production with one digester varies between 37.05 m^3 (July) and 41.83 m^3 (February). With two digesters, production is double, varying between 74.11 m^3 and 83.65 m^3.

In the months of highest production (high average temperatures), January, February, November, and December, the average temperature exceeds 16.4°C, which favors greater microbial activity in the digester. Biogas production in one digester remains above 41 m^3.

In the months of lowest production (low average temperatures), June and July, the average temperatures are the lowest (~14.75 °C to 15.00 °C), and biogas production decreases to 37.05 m^3 in one digester.

The impact of higher temperatures favors microbial activity, which increases biogas production. At lower temperatures, microorganisms slow down their activity, reducing anaerobic digestion. Therefore, using two digesters doubles the production capacity. This is consistent and proportional in all months.

There is a direct relationship between the average monthly temperature and biogas production. The months with warmer temperatures maximize production, and there is a need to double the number of digesters and increase production proportionally. To ensure that this production is achieved through improvement strategies, the digesters must be thermally insulated or heating systems implemented to stabilize production during cold months (**Figure 6**).

The interpretation of the trend lines presents in the function of biogas production with a biodigester, an equation: $y = 2.5121x - 2E{-}12$ and presents a lower slope 2.5121, which implies that the biogas production using a single biodigester increases linearly with x, but at a slower rate. The biogas production line with two biodigesters

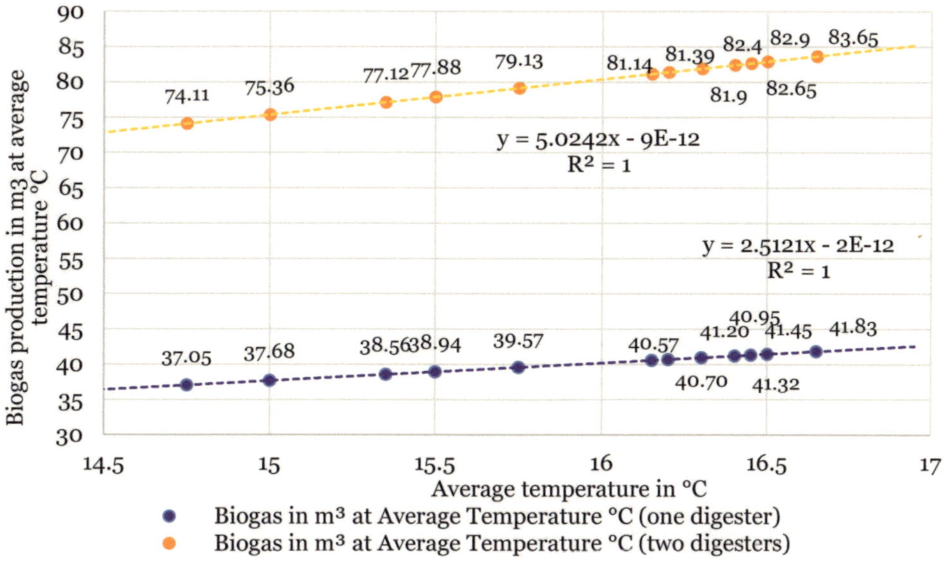

Figure 6.
Biogas production by monthly temperature in Arequipa.

represented by the equation $y = 5.0242x - 9E{-}12$, presents a greater slope 5.0242, which is exactly double the slope of the first and expresses that the biogas production using two biodigesters also increases linearly with x, but at twice the rate than with a single biodigester. It simply means having greater availability of biogas.

There is a direct proportionality, the slope of the line of the two biodigesters 5.0242 is exactly twice that of the slope of the line of one biodigester 2.5121. This indicates that biogas production scales proportionally to the number of biodigesters. Both lines maintain strict linear behavior, reflecting that there is no loss of efficiency when using an additional biodigester. The relationship is completely proportional.

The coefficient of determination R2 = 1 indicates that the trend lines perfectly fit the observed data. Biogas production in both cases (one biodigester and two biodigesters) follows an exact linear relationship with respect to the variable x. There are no variations or discrepancies in the data that explain the relationship, and this may be due to the sample of inhabitants living in the buildings.

The perfect linear efficiency of both systems (one biodigester and two biodigesters) is perfectly proportional in biogas production, doubling production when the number of biodigesters is doubled. R2 = 1 suggests an ideal model, which might not consider possible efficiency losses, technical problems, or external effects in practical applications, or it may be that as the number of inhabitants increases, this value is altered. However, this model can be useful to estimate the biogas production from multiple biodigesters in an ideal context.

The diagram illustrates the behavior of a compact biodigester throughout the year, considering variables such as biogas production, environmental factors, and energy demand, considering variables such as biogas production with one digester, with two digesters, environmental factors, the expected biogas need is identified.

Elements that intervene in the diagram are identified, on the X axis months of the year represents the 12 months of the year, indicating the seasonal behavior of the biodigester, on the Y axis presents the biogas production and other parameters (**Figure 7**).

The blue curve represents production with a biodigester under natural conditions, while the yellow curve simulates ideal conditions with thermal adjustments. The red line marks the energy demand necessary to meet consumption.

The blue curve of production with a biodigester presents a cubic polynomial model: y = 0.0074x3–0.0294x2–0.804x + 43.063, with a coefficient of determination R2 of 0.6988, indicating a moderate fit. They represent the actual biogas production under natural conditions of a biodigester, and it is directly affected by the actual average temperature (dark pink). In cold months, production falls below the purple line (necessary demand), indicating that the system cannot meet demand.

The blue curve of biogas production in two digesters, with the model: y = −0.0148 x3–0.0589x2 + 1.604x + 85.125, also has an R2 fit of 0.6988.

The yellow curve represents an ideal production with fits, it is the biogas production if the ideal conditions (optimal temperature, balanced load) were achieved. It has two biodigesters and is influenced by the light pink bar (ideal temperature). This curve stays closer or above the purple line, showing that reaching the ideal calculated temperature can guarantee the necessary biogas supply.

The red horizontal line represents the expected biogas production and represents the amount of biogas needed to meet the energy demand of all homes, which is a target value that the biodigester system must reach or exceed each month.

The production curve fluctuates around this line. In some months (especially the colder ones), production falls below the expected value, indicating insufficiency in covering demand when biogas production is used with a biodigester.

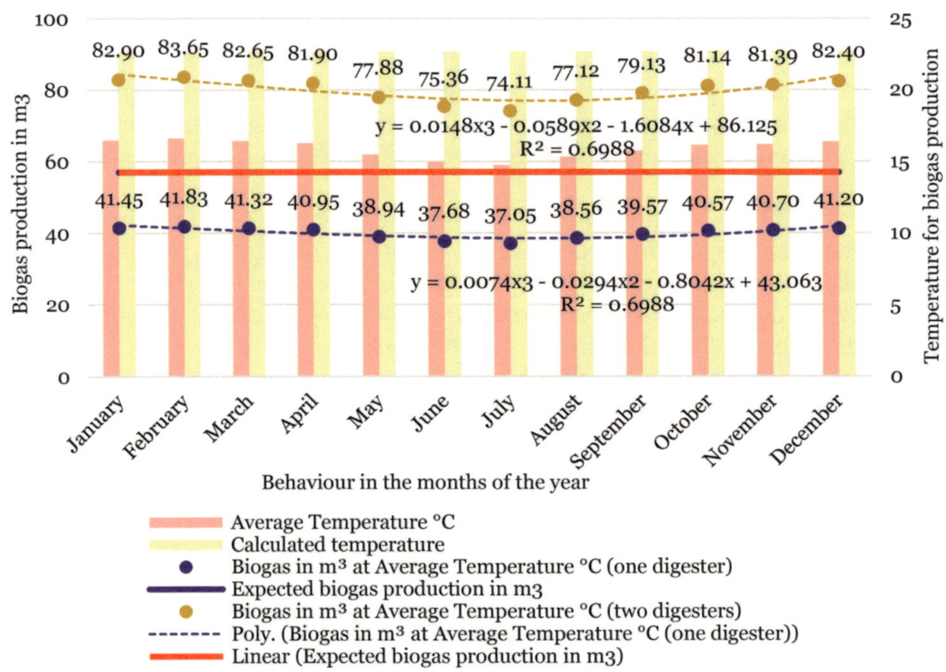

Figure 7.
Biogas production during the year as a function of temperature.

The dark pink bar represents the actual average temperature that occurs in the city of Arequipa and represents the actual temperatures of the environment where the biodigester operates. These natural conditions directly influence microbial activity and, therefore, biogas production. In the cold months, the actual average temperature is low, which significantly reduces production.

The light pink bar represents the ideal calculated temperature, which shows the optimal temperature required to maximize biogas production. This is an indicator of the gap between actual and ideal conditions. This reflects that, in most months, the actual temperature does not reach the ideal temperature, which explains the biogas insufficiency in some periods.

4.2 Compact biodigester efficiency

The efficiency of the compact biodigester reflects the capacity of the system to transform organic waste into biogas and valuable byproducts, such as fertilizers, under optimal operating conditions. It represents the relationship between the amount of energy or biogas produced and the amount of raw material used, considering factors such as design, size and operation of the biodigester. Although they take up less space, compact biodigesters are often more efficient in urban environments thanks to their optimized design to control operating conditions.

The efficiency of the compact biodigester ensures that small amounts of household waste are sufficient to cover basic energy needs, such as cooking. It features a sustainable design that can be used as an ideal solution to reduce the environmental impact of organic waste and minimize dependence on fossil fuels.

The compact design of biodigesters offers several benefits, especially in environments with limitations in space, costs, and operational efficiency. Compact biodigesters are designed to take up less space, making them ideal for urban, residential environments, or areas with land limitations and can be installed in small spaces such as small yards, roofs, or even underground, adapting to urban needs. Their smaller volumetric capacity facilitates the control of internal parameters such as temperature and pH, optimizing biogas production., which maximizes biogas production. Their compact design facilitates thermal insulation, which reduces energy losses and improves anaerobic digestion. Fewer materials are required for their construction compared to larger systems. The small size facilitates cleaning, inspection, and repair tasks, reducing operating costs.

They are ideal for processing organic waste generated in homes, such as food scraps or garden waste. They generate enough biogas to cover basic energy needs such as for cooking, heating, lighting.

Despite their many advantages, compact biodigesters also face certain limitations, such as a lower capacity to process large volumes of waste and being more sensitive to extreme variations in operating conditions, such as temperature and excessive loads.

The compact design of a biodigester is an ideal solution for urban environments or areas with space restrictions. They offer high efficiency, ease of operation, and reduced costs, making them a viable option for households, small businesses, and communities looking to use organic waste sustainably (see **Figure 8**).

This diagram presents an analysis of biogas production with one digester and two digesters over a whole year, comparing it with the specific biogas needs of different family homes (VUF).

The X-axis represents the months of the year (January to December), indicating the seasonal behavior. The Y axis (Left) represents the biogas production in m3,

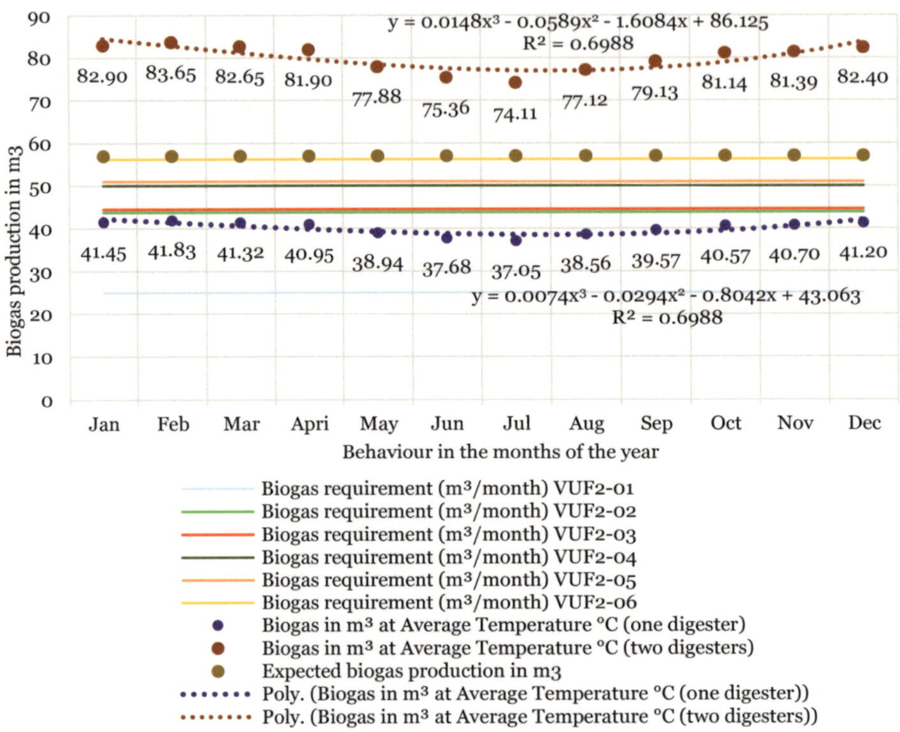

Figure 8.
Efficiency of biogas production during the year depending on temperature.

which varies according to the conditions of the digester and the specific demands. The Y axis (Right), expressed in m3, refers to the polynomial lines that describe the trend of biogas production.

The orange curve represents production with two digesters, highlighting a higher biogas production compared to a single digester. It has a cubic polynomial trend described by the equation: y = 0.0148x3–0.0589x2–1.6084x + 86.125. It shows that production is higher than that of a digester and follows a downward trend during the coldest months (May to July). It recovers from August and exceeds 80 m^3 in December. The R2 = 0.6988 indicates a moderate fit.

The blue curve and blue markers represent the production with one digester and express the real biogas production with one digester. It has a cubic polynomial trend described by the equation: y = 0.0074x3–0.0294x2–0.8042x + 43.063. Production is approximately half of the biogas production with two digesters. It reaches a minimum in July ~37 m3/month and recovers in the warmest months such as in December ~41 m3/month. The model fit is also moderate R2 = 0.6988.

In a temporal analysis, it indicates that during the cold months (May, June, and July), lower temperatures affect microbial activity, significantly reducing biogas production. The blue and orange curves drop significantly due to the decrease in average temperature, which affects microbial activity. In the warm months (December, January, and February), higher temperatures favor anaerobic digestion, increasing biogas production.

Compact biodigester represents an efficient and sustainable solution for urban environments, offering high efficiency, ease of handling, and reduced operating costs. Its design responds to the energy needs of homes while minimizing environmental impact. With one digester, the minimum biogas demands of the smallest homes (VUF2-01 to VUF2-03) are met, while two digesters allow the needs of even larger homes to be exceeded.

4.3 Energy cost reduction

The transition from LPG to biogas in urban homes offers a key opportunity to reduce energy costs, improve environmental sustainability, and promote energy self-sufficiency. This section analyses the economic and environmental benefits of biogas, highlighting its contribution to a more sustainable energy model.

LPG consumption in urban homes in Arequipa.

In Arequipa, LPG is mainly used for cooking food, with an average monthly consumption of 1.06 10 kg cylinders per family at a cost of S/60.00 per cylinder. This consumption generates an average monthly expense that varies between S/79.44 and S/178.74, depending on the size of the family unit. This expense represents a significant part of the family budget, especially for large families.

Energy cost comparison between LPG and biogas.

Table 6 illustrates LPG consumption, its cost, and the impact of replacing it with biogas in different types of urban family homes:

Replacing LPG with biogas eliminates its associated costs, generating significant savings ranging from S/953.28 for a family of 4 to S/2144.88 for a family of 9. The initial investment of S/2000 per biodigester can be quickly amortized thanks to the annual savings in LPG costs.

Sustainable waste management is for biogas production from organic waste, which transforms waste into useful energy, significantly reducing its environmental impact. Biogas captures and reuses methane, a potent greenhouse gas, that would otherwise be released in landfills. Biogas harnesses methane, a potent greenhouse gas, that would otherwise be released in landfills. The digestate, a byproduct of anaerobic digestion, can be used as an organic fertilizer, contributing to the circular economy.

The biodigester designed for urban homes in Arequipa is compact, efficient, and economical. The main materials include PVC or HDPE for the chambers, for their durability and resistance to corrosion. The steel frame ensures structural support, while 4-mm polycarbonate improves thermal insulation. Mixing paddles facilitates the agitation of the biomass to optimize anaerobic digestion. Consider the control

Home	Inhabitants	Biogas (m^3/month)	LPG consumption (kg)	Monthly LPG Cost (S/)	Annual LPG Cost (S/)
VUF_2-01	4	25	13.24	79.44	953.28
VUF_2-02	7	43.75	23.17	139.02	1668.24
VUF_2-03	7	43.75	23.17	139.02	1668.24
VUF_2-04	8	50	26.48	158.88	1906.56
VUF_2-05	8	50	26.48	158.88	1906.56
VUF_2-06	9	56.25	29.79	178.74	2144.88

Table 6.
Comparison of energy costs between LPG and biogas in urban housing.

valves to regulate the flow of biogas and organic materials. The estimated cost is S/2000 per unit, pending adjustments and optimization.

Replacing LPG with biogas in urban homes is an economically viable and environmentally sustainable solution. Implementing compact biodigesters favors the sustainable management of waste, transforming it into useful energy and fertilizers, which results in significant annual savings compared to the use of LPG. It allows waste to be managed sustainably, valorizing organic waste into energy and fertilizers.

It contributes to mitigating environmental impacts by reducing methane emissions and the carbon footprint.

4.4 Environmental impact of biogas production and organic solid waste management

Biogas production in family homes generates both individual and collective benefits, promoting urban sustainability and mitigating key problems in cities, influencing the negative impacts of a city, which affects the following:

Reduction of urban solid waste, homes that produce biogas transform organic waste (food scraps, plant matter, etc.) into energy, reducing the amount sent to the municipal collection system or landfills. It affects the lower pressure on collecting and final disposal services and at the same time contributes to the reduction of the accumulation of waste in landfills, which prolongs their useful life, and this action reduces the generation of leachates and greenhouse gases, prolonging the useful life of landfills.

Mitigation of climate change, the methane generated in the biodigester, a greenhouse gas 25 times more potent than CO_2, is captured and used, preventing its release into the atmosphere. This contributes to reducing the carbon footprint of the home and this is added to the city. This encourages the adoption of low-carbon technologies and contributes to meeting global climate goals. The capture and use of methane to generate energy contributes to reducing the carbon footprint of homes and mitigating the effects of climate change. Studies such as Surendra et al. [67] and the UN [64] highlight that technologies such as biodigesters can reduce GHG emissions by up to 30% and extend the life of landfills.

Reduction of dependence on non-renewable energy, the biogas generated can replace non-renewable energy sources such as LPG, electricity, or firewood for cooking and heating and contributes to influencing the lower demand for fossil fuels, reducing costs and emissions associated with their transport and use. In addition, it reduces the falling of trees for firewood, preserving urban and peri-urban green areas.

Reduction of soil and water pollution, the treatment of organic waste in biodigesters prevents its accumulation in the open air, reducing the contamination of soils and bodies of water. This improves urban environmental quality, reducing the risk of diseases associated with poor waste disposal.

Production of biofertilizers, refers to digestate, as a byproduct of the biodigester, can be used as a natural fertilizer in gardens or family crops. This encourages urban agriculture, reduces the need for chemical fertilizers, and improves local food sustainability. Use in city parks and gardens reduces the need for chemical fertilizers, decreasing the environmental impact of their production and use.

Promotion of the circular economy, biogas production is aligned with the principles of the circular economy, transforming waste into resources such as energy and fertilizers and at the same time generates a benefit for the city and this encourages sustainable management of urban resources and relieves pressure on municipal waste management systems. The compact biodigester reinforces a circular economic approach by transforming organic waste into energy resources and fertilizers. This approach is consistent with the sustainability principles promoted by the Ellen MacArthur Foundation [68].

Cost reduction for the city, the reduction of waste and lower consumption of conventional energy reduce municipal costs associated with transportation, treatment and final disposal. This is a benefit for the city, since it affects the reduction of municipal costs in transportation, treatment, and final disposal of solid waste and, in addition, alleviates energy demand, minimizing the need to expand urban infrastructure.

Economic impact and cost reduction, according to the economic analysis, shows that replacing LPG with biogas can significantly reduce household energy costs, contributing to energy self-sufficiency and climate change mitigation. Cost reduction goes beyond direct savings; it also includes benefits derived from efficient waste management and biofertilizer production. The results show that the initial cost of the compact biodigester can be amortized in a reasonable period, depending on the level of use and access to subsidies. This reinforces findings such as those of IRENA [69], which identify biogas as a viable option to reduce dependence on fossil fuels in urban environments.

Environmental awareness and citizen participation, the implementation of biodigesters promotes responsible habits in the separation and use of waste. It is a benefit for the city. This increases environmental awareness and encourages citizen participation in urban sustainability initiatives.

Biogas production in family homes reduces urban problems related to waste management, energy demand, and environmental pollution. While the benefits for a single home are the modest, the mass adoption of this technology could generate a significant change toward a more sustainable and resilient urban model.

4.5 Explore the adoption and scalability of biogas technology

The adoption and scalability of biogas production technology are key challenges for its massive implementation, especially in urban environments. Both concepts are interrelated and reflect the potential of biogas as a sustainable energy solution.

Scalability refers to the capacity of technology to adapt to a greater number of users or contexts, while maintaining efficiency and economic viability. In the case of biogas, scalability requires designs adaptable to different housing sizes and population densities; economic viability that allows affordable costs on a large scale; and modular designs, such as compact biodigesters, that facilitate technological expansion. In addition, economic viability expressed in the production and adoption of biodigesters must be economically accessible on a large scale, preventing costs from growing disproportionately. Modular designs present solutions such as compact and modular biodigesters that facilitate technological expansion in limited spaces. Mass production is linked to the industrial development of biodigesters and specific components, reducing costs and accelerating their adoption.

The adoption of this technology depends on its acceptance and integration into the daily life of families and communities. Factors such as the perception of clear benefits (energy savings and cost reduction), ease of use, available technical support, and cultural compatibility directly influence adoption. In addition, biodigesters should be intuitive to use and require minimal maintenance. They should have technical support: The availability of technical support and spare parts ensures continued use. Initial costs are an initial investment that can be a barrier for many families, underlining the need for subsidies or accessible financing. There must be cultural compatibility; in some cases, there are cultural prejudices or resistance to change that hinder adoption.

Despite these benefits, the adoption and scalability of compact biodigesters face challenges. Key barriers include initial installation costs, lack of technical knowledge on the part of users, and negative cultural perceptions associated with the use of organic waste. To overcome these limitations, it is essential to implement strategies that include economic incentives, community awareness campaigns, and public policies that support the use of renewable technologies. Research such as that by Wellinger et al. [70] and IRENA [71] underscores the importance of combining technical and social approaches to ensure successful implementation.

The analysis carried out identifies the following main barriers: socioeconomics due to high initial costs, lack of financing, and perception of cheaper energy alternatives (such as subsidized LPG). Cultural barriers are expressed in stigmas associated with organic waste management and preference for traditional methods such as the use of firewood or charcoal. Existing infrastructure where there are limited spaces in urban areas limits the installation of biodigesters. Education expressed in low levels of technical and environmental knowledge hinders the acceptance and maintenance of the biodigester.

The availability of waste from families with low volumes of organic waste may make the biodigester unprofitable.

The adoption and scalability of biogas technology in urban environments requires a multidimensional approach that addresses technical, cultural and economic barriers. The design of more compact biodigesters, centralized waste processing, and the implementation of economic and educational incentives can accelerate the transition to more sustainable and resilient energy systems. This model not only improves efficiency in waste management but also positions biogas as a viable and desirable solution in the urban context.

5. Conclusions

The work demonstrates that the implementation of compact biodigesters in urban homes can effectively meet the stated objectives, such as energy self-sufficiency and sustainable waste management. The results confirm the initial hypotheses, showing that it is feasible to produce enough biogas for the energy needs of an average family, in addition to significantly reducing energy costs and environmental impact. This underlines the transformative potential of this technology to promote sustainability in urban environments.

It was shown that biogas production varies proportionally with the size and waste generation capacity of homes. Larger families, which generate larger volumes of organic waste, achieve greater biogas production, optimizing biodigester performance. This finding is consistent with previous studies that highlight the direct relationship between waste generation and energy production. In addition, it is emphasized that biogas technology is adaptable to various urban configurations, allowing its scalability.

The compact biodigester designed in this study has proven to be an efficient and viable technology for biogas production in urban environments. Its optimized design allows for maximizing biogas production in limited spaces, which is crucial in densely populated urban areas. The incorporation of innovations such as mechanical rotating blades has significantly improved the digestion process, by avoiding sedimentation of waste and ensuring a homogeneous mixture of biomass. These features not only improve the energy performance of the system, but also reduce maintenance requirements, positioning the compact biodigester as an ideal solution for environments with space restrictions.

The transition from LPG to biogas offers significant savings in family energy costs, with annual reductions varying between S/953 and S/2144, depending on the size of the family and the amount of biogas produced. This finding underlines the positive economic impact of biogas use, especially in contexts where the cost of LPG represents a considerable economic burden for households. In addition, biogas promotes energy self-sufficiency, decreasing the dependence on fossil fuels and the costs associated with their acquisition and transportation.

Biogas production has a positive environmental impact by contributing to the reduction of solid waste in landfills, which in turn decreases the emission of greenhouse gases, such as methane, and the generation of polluting leachates. In addition, digestate, a byproduct of the biodigester, can be used as organic fertilizer, promoting sustainable agricultural practices in urban and peri-urban areas. This approach not only improves environmental sustainability at the individual level, but also contributes significantly to mitigating environmental problems in urban environments.

The need for biogas in urban dwellings is mainly determined by the number of inhabitants and, to a lesser extent, by the built surface. The results show that the proper design and sizing of biodigesters, considering the density of occupancy of homes, can efficiently meet the energy demands of homes of different sizes. This reinforces the importance of planning energy systems based on real needs and not only on physical characteristics of homes.

Although compact biodigester technology has proven to be technically viable, its mass adoption faces economic, cultural and educational barriers. Initial costs represent a significant challenge for many families, especially in low-income communities. In addition, the lack of technical knowledge about the operation and maintenance of biodigesters, together with cultural prejudices associated with the use of organic waste, limits their acceptance. These barriers need to be addressed through innovative strategies and educational programs.

To facilitate the adoption and scalability of biogas technology, the following is proposed:

- Design of ultra-compact and automated biodigesters: Adapted for integration in small urban spaces and with systems that minimize manual intervention.

- Centralized waste collection and processing systems: These allow families to deliver their unprocessed waste and receive inputs ready to feed the biodigesters. This simplifies the operation and increases the efficiency of the system. Collecting digestate also involves adding value and introducing the exchange process.

- Educational and training programs: These reinforce confidence in technology and promote its cultural acceptance, especially in communities with less familiarity with this type of solution.

The biogas model is framed within the principles of the circular economy, taking advantage of organic waste to generate energy and fertilizers. This approach connects markets, households, and urban waste management systems, promoting a sustainable cycle. Reusing digestate as organic fertilizer not only reduces dependence on chemicals but also boosts urban agriculture, encouraging greater food and energy self-sufficiency.

The use of biogas has the potential to transform urban sustainability by simultaneously addressing problems related to waste management, energy dependence, and climate change. Its massive implementation in urban housing can significantly contribute to reducing greenhouse gas emissions, promoting sustainable practices and reducing pressure on solid waste management systems. This positions biogas as a key solution to the environmental and energy challenges of cities.

The incorporation of biodigesters into sustainable urban policies can optimize waste management and foster energy self-sufficiency in cities. The results of this study suggest that the adoption of this technology, complemented by economic and educational incentives, can transform cities into resilient and sustainable systems, aligning global sustainability and climate change mitigation goals.

This study provides a solid foundation for future research on:

- The development of more compact and efficient automatic biodigesters.
- The creation of business models based on the circular economy.
- The exploration of educational and financing strategies to promote the mass adoption of technology.
- The analysis of the long-term environmental impact and the integration of these systems into public policies at local and national levels.

These findings highlight the transformative impact of biogas technology in urban housing, laying the foundation for its large-scale implementation and promoting more sustainable and resilient urban development.

Conflict of interest

The authors declare no conflict of interest.

Author details

Guillermo Yorel Noriega Aquise
Catholic University of Santa Maria, Peru

*Address all correspondence to: gmonoriegae@gmail.com; gnoriega@ucsm.edu.pe

References

[1] Arela-Bobadilla R, Riesco-Lind G, Chávez G. Una Mirada a la Expansión de la Ciudad de Arequipa en los Últimos 40 Años. Informe del Centro de Estudios en Economía y Empresa. Arequipa, Peru: Universidad Catolica San Pablo; 2021. Available from: https://ucsp.edu.pe/archivos/publicaciones/centro-estudios-economia-empresas/Expansion-de-la-ciudad-de-Arequipa-CEE-UCSP_2021-09-29.pdf

[2] OEFA Organismo de Fiscalización Ambiental. La fiscalización ambiental de Aguas Residuales. Ministerio del Ambiente Gobierno del Perú. Comité de redacción Equipo de subdirección de supervisión y Equipo de la oficina de comunicaciones. Lima, Perú; 2014. Available from: https://sinia.minam.gob.pe/sites/default/files/sial-sialtrujillo/archivos/public/docs/3171.pdf

[3] O. O. d. E. y. F. Ambiental. Fiscalización ambiental en aguas residuales. Equipo de la Oficina de Comunicaciones y Atención al Ciudadano 2014. Available from: https://hdl.handle.net/20.500.12788/287

[4] Weiland P. Biogas production: Current state and perspectives. Applied Microbiology and Biotechnology. 2010;**85**:849-860

[5] Holm-Nielsen JB, Al Seadi T, Oleskowicz-Popiel P. The future of anaerobic digestion and biogas utilization. Bioresource Technology. 2009;**100**(22):5478-5484

[6] Olgyay V. Design with Climate: Bioclimatic Approach to Architectural Regionalism. EEUU: Princeton University Press; 2015. DOI: 10.2307/j.ctvc77kqb. Available from: https://www.jstor.org/stable/j.ctvc77kqb. EISBN 978-1-4008-7368-5

[7] Gonzalo GE. Manual de Arquitectura Bioclimatica y Sustentable. Universidad Nacional de Tucuman Buenos Aires Argentina. CEEMA; 2019. Available from: https://www.academia.edu/41191010/GEGonzalo_Manual_Arquitectura_Bioclimatica. ISBN 950-43-9028-5

[8] Arthurson V. Closing the global energy and nutrient cycles through application of biogas residue to agricultural land–potential benefits and drawbacks. Energies. 2009;**2**(2):226-242

[9] Balaras CA, Droutsa K, Dascalaki E, Kontoyiannidis S. Heating energy consumption and resulting environmental impact of European apartment buildings. Energy and Buildings. 2005;**37**(5):429-442

[10] Gómez R, Flores F. Ciudades sostenibles y gestión de residuos sólidos. Agenda. Propuestas para mejorar la descentralización. Lima, Peru: Universidad del Pacifico, Centro de Investigacion; 2014. Available from: https://repositorio.up.edu.pe/backend/api/core/bitstreams/6963d42b-8eae-4e32-820f-c33271c59bd7/content

[11] Graziani P. Economía circular e innovación tecnológica en residuos sólidos: Oportunidades en América Latina. America Latina y el Caribe: Caracas: CAF Corporación Andina de Fomento; 2018. Available from: https://scioteca.caf.com/handle/123456789/1247. ISBN 978-980-422-092-0

[12] Patel V, Pandit S, Chandrasekhar K. Basics of methanogenesis in anaerobic digester. In: Kalia VC, editor. Microbial Applications Vol. 2: Biomedicine, Agriculture and Industry. Indian Institute

of Technology Kharagpu, Sharda University, Indian Institute of Chemical Technology. Springer International Publishing AG; Apr 2017. pp. 291-314. DOI: 10.1007/978-3-319-52669-0_16

[13] Chen L, Neibling H. Anaerobic Digestion Basics. Vol. 6. EEUU: Agricultural Experiment & UI Extension Publications, University of Idaho Library Digital Collections; 2014. Available from: https://www.lib.uidaho.edu/digital/uiext/items/uiext33168.html

[14] Ochieng R, Gebremedhin A, Sarker S. Integration of waste to bioenergy conversion systems: A critical review. Energies. 2022;**15**(7):2697

[15] HomeBiogas. One home's trash is another home's treasure. Available from: https://www.homebiogas.com/solutions/backyard/ [Accessed: July 2024]

[16] A. B. Council. Harness the benefits of biogas. Available from: https://americanbiogascouncil.org/resources/why-biogas/ [Accessed: July 2024]

[17] Capareda S. Bioenergy conversion systems. In: Holden NM, Wolfe ML, Ogejo JA, Cummins EJ, editors. Introduction to Biosystems Engineering. American Society of Agricultural and Biological Engineers (ASABE) in Association with Virginia Tech Publishing. Virginia EEUU: Virginia Tech Libraries; 2020. DOI: 10.21061/IntroBiosystemsEngineering/Bioenergy_Systems

[18] Verner V et al. Economics and perception of small-scale biogas plant benefits installed among peri-urban and rural areas in Central Vietnam. Biomass Conversion and Biorefinery. 2023;**13**(13):11959-11971. DOI: 10.1007/s13399-021-02122-4

[19] Muvhiiwa R, Hildebrandt D, Chimwani N, Ngubevana L, Matambo T. The impact and challenges of sustainable biogas implementation: Moving towards a bio-based economy. Energy, Sustainability and Society. 2017;**7**(1):20. DOI: 10.1186/s13705-017-0122-3

[20] H. Biogas. How to make biogas at home—A step-by-step guide. Available from: https://www.homebiogas.com/blog/how-to-make-biogas-at-home/

[21] A. C. D. A. FACT Foundation. Manual for the construction and operation of small and medium size biogas systems

[22] Energy F. Feedstocks for Biogas. Available from: https://farm-energy.extension.org/feedstocks-for-biogas/

[23] Korbag I, Omer SMS, Boghazala H, Abusasiyah MAA. Recent Advances of Biogas Production and Future Perspective. United Kingdom: IntechOpen; 2020

[24] Atelge M et al. Biogas production from organic waste: Recent progress and perspectives. Waste and Biomass Valorization. 2020;**11**:1019-1040

[25] Beschkov V. Biogas production: Evaluation and possible applications. In: Biogas—Basics, Integrated Approaches, and Case Studies. IntechOpen; 2021. DOI: 10.5772/intechopen.101544. ISBN: 978-1-80355-109-8

[26] Atelge MR et al. A critical overview of the state-of-the-art methods for biogas purification and utilization processes. Sustainability. 2021;**13**:20. DOI: 10.3390/su132011515

[27] Abanades S et al. A critical review of biogas production and usage with legislations framework across the globe. International journal of Environmental

Science and Technology. 2022;**19**(4):3377-3400. DOI: 10.1007/s13762-021-03301-6

[28] India E. E. A. Bio-CNG production—Technology and process details. Available from: https://www.eai.in/blog/biobiz/2020/03/bio-cng-production-technology-and-process-details.html

[29] C. Making. Biogas Production Equipment. Available from: https://making.com/biogas

[30] I. IRENA. Measuring Small-Scale Biogas Capacity and Production. United Arab Emirates: IRENA Abu Dhabi; 2016

[31] Mahmoodi P, Farmanbordar S, Karimi K. Analytical methods in Biogas production. In: Tabatabaei M, Ghanavati H, editors. Biogas: Fundamentals, Process, and Operation. Cham: Springer International Publishing; 2018. pp. 221-238

[32] R. Global. Building a more sustainable future. Available from: https://www.repsol.com/en/energy-and-the-future/future-of-the-world/bioclimatic-architecture/index.cshtml

[33] Yentekakis IV, Goula G. Biogas management: Advanced utilization for production of renewable energy and added-value chemicals. Frontiers in Environmental Science. 2017;**5**:Article 7. DOI: 10.3389/fenvs.2017.00007

[34] Heiker M, Kraume M, Mertins A, Wawer T, Rosenberger S. Biogas plants in renewable energy systems—A systematic review of modeling approaches of Biogas production. Applied Sciences. 2021;**11**(8):3361 [Online]. Available from: https://www.mdpi.com/2076-3417/11/8/3361

[35] Camirand BWE. Biogas plant development handbook. [Online]. Available from: https://www.biogasworld.com/biogas-plant-development-handbook/

[36] Salvi O, Chaubet C, Evanno S. Improving the safety of biogas production in Europe. Revista de Ingeniería. 2012;**37**:57-65

[37] Huber S. Small-Scale Biogas Production from Organic Waste and Application in Mid-Income Countries–A Case Study of a Lebanese Community. Sweden: Uppsala University; 2019. Available from: https://www.diva-portal.org/smash/get/diva2:1334609/FULLTEXT01.pdf

[38] Capodaglio AG, Callegari A, Lopez MV. European framework for the diffusion of biogas uses: Emerging technologies, acceptance, incentive strategies, and institutional-regulatory support. Sustainability. 2016;**8**(4):298 [Online]. Available from: https://www.mdpi.com/2071-1050/8/4/298

[39] G. B. Asociation. Guidelines for the Safe Use of Biogas Technology (Safety First). Alemania: GIZ, Gesellschaft für Internationale Zusammenarbeit y German Biogas Association; 2017

[40] Elashry ER, Ahmed AE, Dorgham HK, Youssef RM. Design and installation of a biogas production unit operating with automatic control system. Misr Journal of Agricultural Engineering. 2024;**41**(1):15-28

[41] Zamorano M, Perez JIP, Paves IA, Ridao AR. Study of the energy potential of the biogas produced by an urban waste landfill in southern Spain. Renewable and Sustainable Energy Reviews. 2007;**11**(5):909-922

[42] Cherubini F, Bargigli S, Ulgiati S. Life cycle assessment of urban waste management: Energy performances and environmental impacts. The case

of Rome, Italy. Waste Management. 2008;**28**(12):2552-2564. DOI: 10.1016/j.wasman.2007.11.011

[43] Paolini V, Petracchini F, Segreto M, Tomassetti L, Naja N, Cecinato A. Environmental impact of biogas: A short review of current knowledge. Journal of Environmental Science and Health, Part A. 2018;**53**(10):899-906

[44] Muhondwa JP, Martienssen M, Burkhardt M. Feasibility of anaerobic digestion of potato peels for biogas as mitigation of greenhouse gases emission potential. International Journal of Environmental Research. 2015;**9**(2)

[45] Vechi NT, Jensen NS, Scheutz C. Methane emissions from five Danish pig farms: Mitigation strategies and inventory estimated emissions. Journal of Environmental Management. 2022;**317**:115319. DOI: 10.1016/j.jenvman.2022.115319

[46] Kuo J, Dow J. Biogas production from anaerobic digestion of food waste and relevant air quality implications. Journal of the Air & Waste Management Association. 2017;**67**(9):1000-1011

[47] Baştabak B, Koçar G. A review of the biogas digestate in agricultural framework. Journal of Material Cycles and Waste Management. 2020;**22**:1318-1327

[48] Lasswell HD. Power and Society: A Framework for Political Inquiry (Politica y Relaciones Internacionales). Pickle Partners Publishing; 2018

[49] Almond GA, Powell GB. Comparative Politics: A Developmental Approach (Little, Brown Series in Comparative Politics). Boston: Little, Brown and Company (in English); 1966

[50] Nohlen D. El método comparativo. Antologías para el estudio y la enseñanza de la ciencia política. 2020;**3**:41-57

[51] Smelser NJ, Swedberg R. Introducing economic sociology. The Handbook of Economic Sociology. 2005;**2**:3-25. Available from: https://citeseerx.ist.psu.edu/document?repid=rep1&type=pdf&doi=7a7f2bdcca886b39d88bb829241f42863570c009#page=16

[52] Nohlen D. Ciencia política comparada: el enfoque histórico-empírico. Granada, España: Editorial Universidad de Granada; 2013. Available from: https://editorial.ugr.es/media/ugr/files/sample-138136.pdf. ISBN: 978-84-338-5584-8

[53] de León GD, de la Garza C y L, de Ayde E. Método comparativo. In: Métodos y técnicas cualitativas y cuantitativas aplicables a la investigación en ciencias sociales. Capitulo 11. Mexico: Tirant Humanidades; 2014. pp. 223-251. Available from: http://eprints.uanl.mx/id/eprint/9943. ISBN 9788416062324

[54] Piovani JI, Krawczyk N. Los Estudios Comparativos: algunas notas históricas, epistemológicas y metodológicas. Educação & Realidade. 2017;**42**(3):821-840

[55] Lijphart A. Comparative politics and the comparative method. American Political Science Review. 1971;**65**(3):682-693. DOI: 10.2307/1955513

[56] Drosg B. Process Monitoring in Biogas Plants. France: IEA bioenergy Paris; 2013

[57] Obileke K, Mamphweli S, Meyer EL, Makaka G, Nwokolo N. Design and fabrication of a plastic biogas digester for the production of biogas from cow dung. Journal of Engineering. 2020;**2020**(1):1848714. DOI: 10.1155/2020/1848714

[58] Vindis P, Mursec B, Rozman C, Janzekovic M, Cus F. Biogas production with the use of mini digester. Journal of Achievements in Materials and Manufacturing Engineering. 2008;**28**(1):99-102

[59] Hafner SD, Rennuit C, Triolo JM, Richards BK. Validation of a simple gravimetric method for measuring biogas production in laboratory experiments. Biomass and Bioenergy. 2015;**83**:297-301

[60] Mittal S, Ahlgren EO, Shukla PR. Barriers to biogas dissemination in India: A review. Energy Policy. 2018;**112**:361-370. DOI: 10.1016/j.enpol.2017.10.027

[61] Abdallah M, Shanableh A, Adghim M, Ghenai C, Saad S. Biogas production from different types of cow manure. In: 2018 Advances in Science and Engineering Technology International Conferences (ASET). United Arab Emirates, Dubai, Sharjah, Abu Dhabi: IEEE; 2018. pp. 1-4. Available from: https://nchr.elsevierpure.com/en/publications/biogas-production-from-different-types-of-cow-manure. ISBN (Electronic)9781538623992

[62] E. INEI. Diccionario de la Encuesta Nacional de Hogares (ENAHO) 2020—Instituto Nacional de Estadística e Informática—INEI. Available from: https://www.datosabiertos.gob.pe/dataset/encuesta-nacional-de-hogares-enaho-2020-instituto-nacional-de-estad%C3%ADstica-e-inform%C3%A1tica-inei

[63] O. GPAE. Informe de Resultados, Consumo y Usos de los Hidrocarburos Líquidos y GLP, Encuesta Residencial de Consumo y Usos de Energía – ERCUE 2019-2020, O. O. S. d. l. I. e. E. y. M.-G. d. P. y. A. E. GPAE, ed., 2022. [Online]. Available from: https://cdn.www.gob.pe/uploads/document/file/2560069/ERCUE%20Hidrocarburos%20%202019-2020.pdf

[64] UN. In: NATIONS U, editor. Updated Guidebook on Biogas Development. 1984. [Online]. Available from: https://repository.unescap.org/bitstream/handle/20.500.12870/6256/ESCAP-1984-MN-Updated-guidebook-biogas-development.pdf?sequence=1&isAllowed=y

[65] V. M. M. T. Manual de biogás. P. d. l. N. U. p. e. D. Proyecto CHI/00/G32. Chile: Remoción de Barreras para la Electrificación Rural con Energías Renovables. Ministerio de Energía, Ed., ed. Organización de las Naciones Unidas para la Alimentación y la Agricultura, Global Environment Facility: MINENERGIA/PNUD/FAO/GEF. Chile: Editado por Proyecto CHI/00/G32 Chile, 2011

[66] Bond T, Templeton MR. History and future of domestic biogas plants in the developing world. Energy for Sustainable Development. 2011;**15**(4):347-354. DOI: 10.1016/j.esd.2011.09.003

[67] Surendra KC, Takara D, Hashimoto AG, Khanal SK. Biogas as a sustainable energy source for developing countries: Opportunities and challenges. Renewable and Sustainable Energy Reviews. 2014;**31**:846-859. DOI: 10.1016/j.rser.2013.12.015

[68] MacArthur F. E. Hacia la economía circular: Razonamiento económico y de negocios para una transición acelerada. Available from: https://www.ellenmacarthurfoundation.org/publications/towards-the-circular-economy-vol-1-an-economic-and-business-rationale-for-an-accelerated-transition

[69] IRENA. Biogas for Domestic Cooking: Technology Brief. Abu Dhabi: International Renewable Energy Agency; 2017. Available from: https://www.irena.org/-/media/Files/IRENA/Agency/

Publication/2017/Dec/IRENA_Biogas_for_domestic_cooking_2017.pdf. ISBN 978-92-9260-039-6

[70] Wellinger A, Murphy JD, Baxter D. The Biogas Handbook: Science, Production and Applications. Philadelphia, PA, EEUU: Woodhead Publishing Limited; 2013. DOI: 10.1533/9780857097415. ISBN 978-0-85709-498-8

[71] I. R. E. A. IRENA. Renewable Power Generation Costs in 2019. 2020

Chapter 3

A Perspective on Biogas Production from Algae Biomass: A Promising Third-Generation Biofuel for the Sustainable Development

Lina Montuori, Manuel Alcázar-Ortega and Abdelhafidh Benreguieg

Abstract

This book chapter explores the biogas production potential of algae and its significance in sustainable energy. The main benefits of algae over traditional biomass sources for biogas generation will be analyzed together with their CO_2 sequestration capabilities, high biomass yield, and rapid growth rates. Moreover, a comprehensive state of the art about biogas production focusing on anaerobic digestion and strategies to enhance biogas yield, putting a particular emphasis on technological innovations, will be presented. Furthermore, a comparative analysis of various production processes, substantiated by experimental data available in the literature, will be carried out. In addition, the possible employment of algae as green fuel for internal combustion engines will be investigated, and the compatibility with current engine technologies will be assessed. Finally, opportunities and constraints to the employment of algae resources on a large scale will be examined, and the policy landscapes that could facilitate widespread of its adoption and integration with other systems will be illustrated.

Keywords: algae biofuel, microalgal biogas, seaweeds, fuel blend, renewables, energy generation, yield, photobioreactor

1. Introduction

Recent studies showcased algae biomass promising role in capturing CO_2 to produce eco-friendly, high-yield biofuels with the potential of building the path to a sustainable energy transition [1, 2]. In the European bioeconomy, algae still represent an untapped resource, and their production accounts for just 0.28% of the global production worldwide. The term 'algae' englobes both microalgae and macroalgae biomass (also called seaweeds). Microalgae are generally cultivated in open ponds or in closed basins, while macroalgae are harvested from wild stocks or produced in aquaculture systems. Aquaculture systems for algae cultivation are still not so common, but they have started to be employed more in recent years [3]. The growing interest in algae biomass as an energy resource resides in algae beneficial properties, mainly the high growth rates, with some species having the ability to double their population in a day under optimum conditions. Microalgae fast development can provide a growth rate of about 50–100 t of dry algae annually per hectare [4], greatly exceeding spread crops used for food, feed, and biofuel production, such as maize and soy, which growth rate ranges between 0 and 20 t annually per hectare [5].

Moreover, they have demonstrated a remarkable capacity to catch carbon dioxide with a potential of up to 1.83 kg of CO_2 for each kilogram of dry biomass generated, providing a twofold benefit in carbon absorption and usage, thereby aiding climate change mitigation efforts [6, 7]. Macroalgae are characterized by unique biochemical properties and ecological benefits that make them suitable for biogas, biobutanol, and bioethanol production. In particular, they represent a third generation of biofuels characterized by a high content of carbohydrates, lipids, and sugars with high efficiency conversion. Despite that, macroalgal-based biofuel production is still not ready to be economically feasible, as barriers exist when the whole production chain is considered [8]. Among them, the complex polysaccharide structures of the macroalgae result in the need for specific and expensive pretreatment methods that negatively impact on the production costs [9]. Substantial water content of algae (sometimes reaching 90%) requires energy-intensive dewatering methods, hence considerably elevating the operating expenses of biogas generation. Moreover, the resilient cell walls of some algae species might impede digestion efficiency, necessitating pretreatment techniques such as mechanical disruption, heat treatment, or chemical hydrolysis to improve biogas production [10].

Economically, algal biogas facilities have high startup costs, but technical advances may lower them [11–13]. To minimize the cost of this technology, strain selection, genetic engineering for biomass quality, and digesting process optimization are being studied [14]. Economic examination of algae as a biogas substrate is complicated by its heterogeneity in production (100–500 mL CH_4/g VS) based on species, growth circumstances, and digestion factors [15]. On the other hand, the seasonal availability of algae biomass is a challenge to ensure a reliable supply. To preserve algae produced in high productive period like summer, where production is generally three to five times higher than winter, drying or ensiling processes can be used. These techniques are especially relevant for applications of algae biomass in the biorefinery sector. The counterpart is that these processes are costly and high-energy-wasting, so that the possible composition changes due to algae biomass storage, an issue that still needs to be fully addressed [16]. Previous studies demonstrated that the ensiling preservation technique based on lactic acid fermentation can stabilize algae biomass (20% solids) and algae-lignocellulosic blends (40% algae/60% lignocellulosic biomass on dry basis) for over 6 months limiting microbial degradation with interesting cost savings and fewer emissions for the biofuel production [17]. Algae are biochemically ideal for anaerobic digestion. On the other side, microalgae contain 20–50% carbohydrates, 10–30% lipids, and 6–71% proteins by dry weight, providing methanogenic bacteria with a large substrate. Therefore, biogas from microalgae is more stable and maybe higher in production with this diversified substrate. Algae cultivation may be included in wastewater treatment systems, where they absorb nutrients such as nitrogen and phosphorus for development, and successfully eliminate them from the water. This bioremediation component can substantially diminish the expenses and ecological consequences of wastewater treatment. Existing studies about algae production, especially microalgae, have been predominantly carried out in laboratories using artificial light in cultivation. But to achieve the maturity of algae carbon capture technologies, long-term studies under natural conditions need to be carried out [18]. In this context, this book chapter explores the potential of algae as a pivotal and sustainable energy source to produce third-generation biofuels. The main benefits of algae over traditional biomass sources for biogas generation will be analyzed together with their substantial CO_2 sequestration capabilities, high biomass yield, and rapid growth rates.

A comprehensive state of the art of microalgal biogas production and techno-economic analysis, with an examination of the current cultivation and production methods, putting a particular emphasis on technological innovations and economic considerations, will be presented. Furthermore, a comparative analysis of various production processes, substantiated by real-world data, will be carried out.

In addition, the possible employment of algae as green fuel for internal combustion engines (ICE) will be investigated, and the compatibility with current engine technologies from algae will be assessed. Finally, opportunities and constraints to the employment of algae resources on a large scale and the policy landscapes that could facilitate widespread of its adoption and integration with other systems will be illustrated.

2. Macroalgae and microalgae biomass as feedstock for biogas production

Algae as feedstock for bioenergy conversion include both macroalgae (seaweeds), characterized by a length of up 60 m, then microalgae, with microscopical cells of an average length between 2 and 200 μm [19].

Asia is leading the market in seaweed production, and countries outside Asia produced less than 2% of the total farmed seaweed volumes in 2020 [20]. Europe, the United States, Australia, and New Zealand are seriously committed to expand the seaweed cultivation, but China and Indonesia are still the main and undiscussed producers [1]. The Chinese government has envisioned a high-technology development plan based on an investment of $3.3 million to develop marine microalgae farm to produce algal biofuel [21]. Developing countries in South Africa, Middle East, and Africa also possess favorable conditions for the large-scale production of algae for biodiesel [22]. In Europe, macroalgae account for an annual production of 31 Mt. wet weight with a harvesting of just 1.2 Mt. [2], with France, Ireland, and Spain as the top 3 countries in number of macroalgae production units [23]. A study carried out by the European Commission mapped 447 algae production units in 23 countries (**Figure 1**). According to the results, more than 50% of these companies produce microalgae, with Germany, Spain, and Italy at the top of the list for microalgae production [24].

2.1 Seaweed cultivation techniques

As already mentioned, seaweed cultivation can be carried out in coastal or off-shore open sea systems, in tanks or ponds. Their cultivation needs to be regulated due to the potential environmental risks of changing the structure of local fauna and flora by introducing parasites or alien species [25]. Due to the absence of lignin, low cellulose content, and the presence of easily fermentable carbohydrates in their composition, marine macroalgae have a higher energy conversion potential of above 100 EJ/yr. If compared with terrestrial biomass [26, 27]. The five main seaweed species groups that are commercially farmed and the main farming countries are shown in **Table 1.**

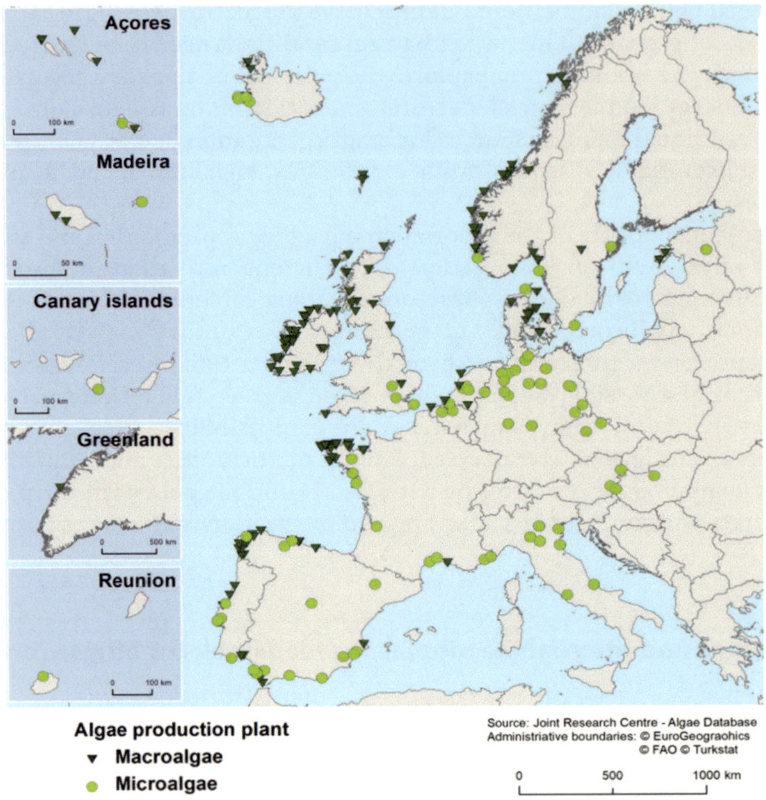

Figure 1.
Location of microalgae and macroalgae production plants (Source: Joint Research Centre – Algae Database Administrative Boundaries EuroGeographics, FAO, Turkstat. Available from: www.ec.europa.eu).

Seaweed family	Seaweed group	Seaweed group species	Farmer countries
Solieriaceae	Eucheumatoids	*Eucheuma seaweeds nei* Spiny Euchema Eikhorn sea Moss	Indonesia Philippines Malaysia
Laminariaceae	Saccharina	*Laminaria japónica* *Saccharina latissima* *Laminaria digitata*	South Korea Japan China
Alariaceae	Undaria	*Undaria pinnatifida*	South Korea Japan China
Bangiaceae	Pyropia	*Pyropia yezoensis* *Porphyra* spp. *Porphyra columbina*	South Korea Japan China
Gracilariaceae	Gracilaria	*Gracilaria gracilis* *Gracilaria verrucosa* *Gracilaria* spp.	Indonesia China

Table 1.
Main seaweed species group that are commercially farmed (Source: Author elaboration from Hatch Innovation Service – Seaweed insights and FAO – Fisheries and Aquaculture).

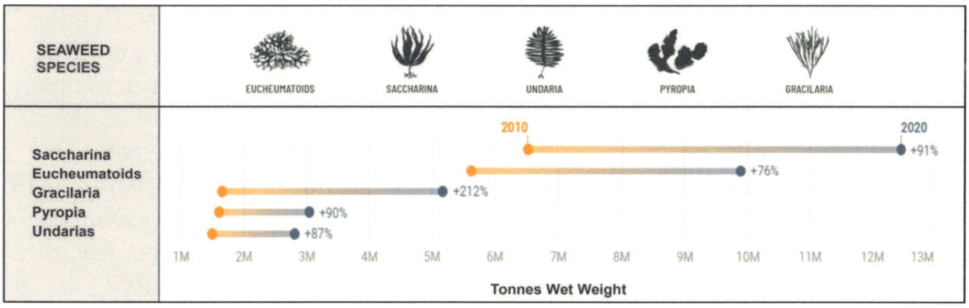

Figure 2.
Growth in production volumes 2010–2020 by species (Source: Author elaboration; Hatch innovation services; FAO fisheries and aquaculture created with datawrapper).

In terms of production volumes, *Gracilaria* species shows the highest growth with a + 212% from 2010 to 2020 (**Figure 2**).

2.2 Microalgae cultivation techniques

Microalgae are photosynthetic organisms that comprise over 3000 aquatic species the majority autotrophic. **Figure 3** shows the rRNA-based phylogeny of some selected microalgae and biogas yield [4].

Microalgae cultivation can be carried out in different ways. Most commonly, they are grown in open raceway ponds, but also advanced technologies are nowadays available, such as closed photobioreactors and two-stage hybrid systems [29]. In the raceway ponds, the cultivation is done in a closed-loop recirculation channel that

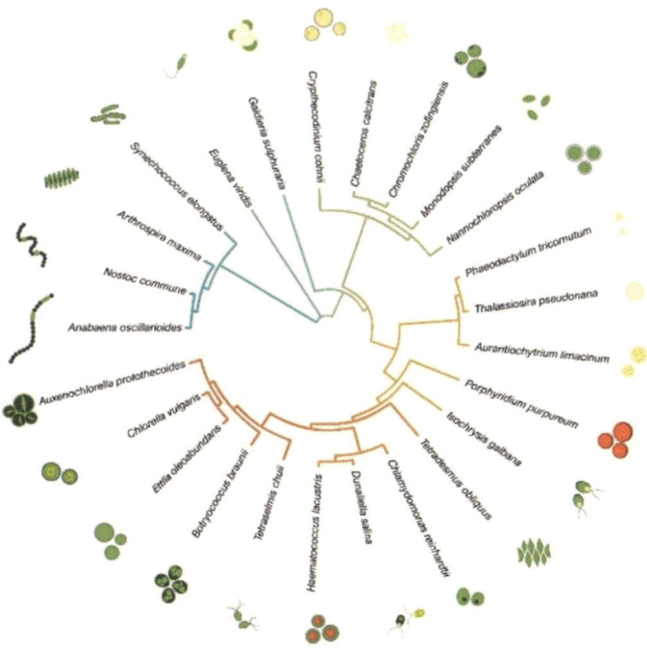

Figure 3.
Phylogenetic tree of 25 industrially relevant microalgal species from GenBank database [28].

can have different lengths and diameters. Open raceway ponds are cost-effective, but they have the disadvantages of high contamination, difficult growth control, and highly variable temperatures. Photobioreactor (PBR) is a competitive microalgae cultivation method where microalgae can grow using artificial light and is able to monitor all input data in real time [30], improving the final biomass density and purity. Moreover, thanks to this technology, mass-scale microalgae cultivation can be optimized by introducing an AI model to control process parameters and large-scale industrial production. The volumetric productivity of microalgal cultivation in PBR is about 0.27–0.56 kg/L/day higher than the volumetric productivity of an open pond, which is about 0.035 kg/L/day [31]. Finally, microalgae cultivation can also be done using two-stage hybrid systems, a technology that combines open ponds with a photobioreactor, reducing the required surface area and that is able to boost the microalgae productivity, cutting the capital and operational costs [32].

3. Anaerobic digestion for algae biogas production

Biomass can be an alternative fuel to address the challenges of sustainable energy supply and reduce fossil fuel dependence. Different biomass conversion technologies are currently available for biogas production from algae. Among them, anaerobic digestion (AD) is a natural biological process in use since 1957 [33]. It is the most direct route for biogas production from algae due to its distinctive advantages. The algae biogas production through AD consists of multiple stages, as shown in **Figure 4**: algae pretreatment, anaerobic digestion, and biogas storage. A series of actions undertaken by various groups of bacteria takes place in a bioreactor, where conversional phases of hydrolysis, acidogenesis, acetogenesis, and methanogenesis occur. The result is the transformation of organic materials into biogas (methane, carbon dioxide) and digestate (bacterial biomass). The digestate, together with the CO_2, is used as a nutrient for the algae growth in the ponds [36].

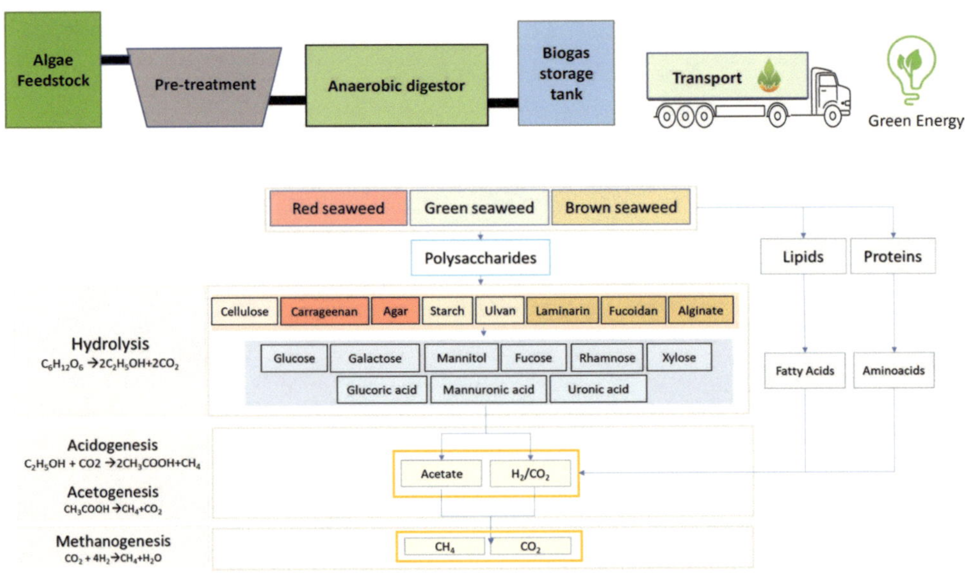

Figure 4.
Seaweed process value chain for biogas production (Own elaboration by adaptation of Refs. [34, 35]).

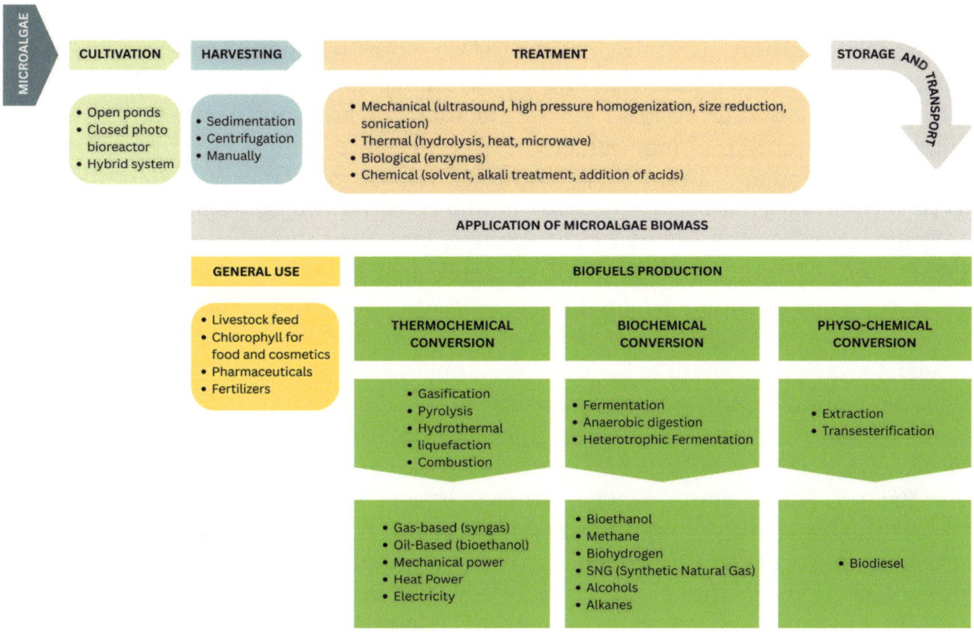

Figure 5.
Microalgae process value chain for biofuels production [39].

Seaweed has been used for biogas production since the nineteenth century. The high content of water and the high productivity rate that characterized seaweed made them preferable to be treated by AD. In fact, biogas production from seaweed is closer to industrial exploitation than biogas production from microalgae [37]. The obtained biogas typically contains methane (50–70%), carbon dioxide (30–45%), hydrogen (<2%), and hydrogen sulfide (<3.5%) [34, 35, 38].

Currently, seaweed is largely used in food, beverage, and cosmetics sectors due to its nutritional properties. Indeed, the high polysaccharide content allows them to absorb CO_2 and pollutants from the wastewater. Concerning microalgae, they are mainly used in commercial applications as livestock feed, bioactive compounds such as antibiotics, or as a nutrient supplement for human consumption. Ultimately, they are being exploited for biofuel production by using different methods. Algal oil can be used for biodiesel production through transesterification and extraction. Thermochemical processes such as gasification, pyrolysis, hydrogenation, liquefaction, and combustion can be used to produce syngas, electricity, or heat power from microalgae. Finally, as depicted in **Figure 5,** biochemical conversion processes such as fermentation can be used, like AD, in order to produce bioethanol, methane, and others [39].

The proper selection of the biogas technology is fundamental for the feasibility and cost-effectiveness of the production process. Depending on the production technology adopted, the biogas yield can vary in a wide range. For instance, in the case of bio-methane, yields could vary from 25% up to 190% [40].

4. Algae biogas applications to internal combustion engines

Under the Climate Change and Energy Transition Law, entered into force on 20 May 2021, new passenger cars and light commercial vehicles must be emission-free, and no new exploitation authorizations will be allocated for hydrocarbon research permits or concessions. Despite that, the EU transport sector still accounts for 60% of the total air pollution due to automobiles running on fossil fuels. In 2023, in Spain, petroleum accounted for roughly 45% of the total consumption of primary energy, while the consumption of renewables (REN) represented only 21%. The total volume of petroleum products domestically consumed was derived from over 30 million metric tons of diesel products; meanwhile, the consumption of gasoline stood at approximately 6.1 million metric tons that year [41]. Additionally, the impact of the COVID-19 pandemic and the geopolitical tensions that led to the Ukraine War highlighted the socio-economic and political vulnerability of the EU and its energy dependence on imported fossil fuels. In terms of energy supply, again in Spain in 2024, the transport sector was shown to be the highest energy-consuming sector, accounting for 38.4% of the final energy demand, dominated by oil products (80.7%), the majority of which are imported (96.1%) [42]. In this scenario, biomass, and consequently biofuels, are at the top of the list of alternatives to petroleum-derived fuels with the potential to promote energy independence to Europe and globally. Algae biofuel can be a sustainable resource to replace traditional fossil fuels due to its biological traits, the compatibility of the existing infrastructure, and its environmental advantages. In this context, research studies need to accelerate the investigation of the adaptability of biofuels to the existing reactor and internal combustion engine (ICE).

Previous studies showed that a co-digested mixture of raw algae and pig manure to be able to produce a higher yield of methane-rich biogas in ICE [43]. Moreover, biogas and biohydrogen blends supplied to biodiesel derived from algae in the ratio of 14:6 emitted the lowest effluents when fueled in diesel ICE, improving combustion efficiency, reducing NOx emission, and particulate matter [44, 45]. The conversion of algal lipids into biodiesel entails transforming them into fatty acid methyl esters (FAMEs), which possess characteristics such as cetane number, viscosity, and cold flow that closely resemble those of petroleum diesel, making them eligible for direct application in diesel engines or blending [46].

Dual-fuel ICE can use two different fuels: a primary fuel for the combustion and an ignition starter as a secondary fuel. ICE operating in dual mode by using biogas and biodiesel has been demonstrated to have better performance and emissions compared with single-fuel engines [47].

Biogas produced from *Chlorella vulgaris* microalgae blended with hydrogen and nanoparticles in diesel, when used in ICE, demonstrates an increase in engine performance with the decrease of the maximum combustion pressure [45]. Until now, *Chlorella vulgaris* microalgae has been used in novel experiments for producing algal biodiesel through a transesterification process, where results have proved that, when adding potassium hydroxide and methanol to the obtained algal biodiesel, the yield of the diesel engine was higher, reaching 85%. But what is more important is that no amendments were required to run the engine with algae biodiesel, and the emission of CO, UBHC, NOx, and smoke opacity was much lower than those of a diesel engine, reaching the maximum decrease of 10 ppm in NOx at a power of 1.19 kW [48]. Likewise, another study carried out on a diesel engine again fueled with *Chlorella vulgaris* microalgae blends and assisted with nanomaterials and biogas rich in HY and CO showed an increase in the ICE performance in the presence of carbon monoxide [49].

Furthermore, alternative approaches to utilizing microalgal biomass in diesel engines have demonstrated that algal slurry diesel emulsions with up to 6.6% wt/wt biomass emit lower exhaust NOx and PM than fossil diesel [50].

Multiple studies have been found on *Chlorella vulgaris* microalgae as a primary source for biofuel production due to such beneficial chemical properties as higher heating values and low kinematic viscosity values when compared to other microalgae [15, 51–53]. Algae spirogyra methyl ester is another microalga that has been often tested in diesel engines. A recent study on a tri-fuel agricultural diesel engine operating algae spirogyra methyl ester showed that a smoother pressure rise is achieved by using algae residual carbon nanoparticles [54]. Moreover, an additional experiment carried out by the same authors assessed that on a dual-fuel combustion engine fueled by HY blended with algae spirogyra biodiesel, these HC, CO, and smoke emissions strongly decreased due to this fuel mix [55]. Although modest variations in fuel efficiency may arise from variances in chemical composition, algal biodiesel is generally regarded as a feasible substitute or supplement for diesel in several internal combustion engine applications. Moreover, engine evaluations have consistently demonstrated that algal biodiesel and biogas blends can sustain and enhance the performance parameters of conventional diesel fuels [56].

Though the positive aspects abound, several challenges remain. The variability of the composition of the raw material species due to differences in the biogas production process is still a barrier to the large-scale application of biogas in ICE. The biogas blend with algae biofuels, and HY shows better performance than a diesel engine, but further studies are required to control the process parameters in detail [57].

The production of algal biofuel is limited by its economic feasibility due to high agricultural, harvesting, and processing costs. Advancements in energy-efficient cultivation and conversion technologies are needed, together with strong quality control on the growing environment, to improve the biofuel properties [58].

Recently, the company Volkswagen launched into the market an algae biogas-powered vehicle. The clean biogas fuel is obtained from microalgae with treated wastewater. The vehicle is equipped with a double fuel tank, one for biogas and one for emergency, and the car can run on compressed natural gas and biogas [59].

Finally, in 2024, Barbados University, in collaboration with The University of West Indies (UWI) Cave Hill, Rum and Sargassum Inc., and Supernova Lab of Future Barbados, presented the first vehicle powered by sargassum seaweed and rum waste to create biofuel [60].

During the latest years, algae biomass has also been seen by the army as a strategic fuel to be used in battle tank engines. The US Army Combat Capabilities Development Command Army Research Laboratory (DEVCOM ARL) and the Indian Army have focused their research since 2019 on biomass-to-energy devices with the goal of reducing the volume of energy supplies that soldiers must carry, such as batteries and electric generators, reducing operation risks and allowing soldiers to come home safely. Ideally, the army can cultivate algal biomass in camps and produce biofuel to be used in military vehicles in order to extend operations during wartime. Indeed, some algae grow in as little as 4–6 hours [61, 62].

5. Strategies for improving biogas production

The production of biogas from microalgae still has some limitations that need to be addressed, such as high maintenance costs, low biomass productivity, and low yield. However, synergies with more mature technologies such as biorefinery could improve their applicability [55]. The effectiveness of the biogas production relies mainly on the optimization of the driving parameters. Among them, temperature, pH, retention time, and waste have been identified as crucial factors able to affect directly the AD biogas yield and stability [63]. An unwanted generation of sulfide, an excess of salinity, and lignin components have been demonstrated to affect negatively biogas production [36]. Process temperature regimes are specific for each algae species and should be kept under control. During the AD process, different possible temperature ranges have been identified: psychrophilic (5–20°C), mesophilic (25–45°C), and thermophilic (45–65°C).

Similarly, the pH varies at each stage, generally ranging between seven and eight for a single-stage AD [64]. **Table 2** shows an example of the effects of different operational parameters on biogas yield for the two selected microalgae at two different temperatures and pH values.

Pretreatment methods such as mechanical, thermal, microwave, enzymatic, and catalytic irradiation can be applied to boost biogas production [13]. Actually, they have the capability to improve biomass solubility by breaking down algae cell walls and increasing the production of lipids and proteins. It has been demonstrated, for example, that microalgae such as Chlorella and Scenedesmus could produce 150–200 mL CH_4/g VS without pretreatment, whereas after thermal or mechanical pretreatment, the yield can increase up to 450–600 mL CH_4/g VS. The significant influence of pretreatment on biogas production is shown in **Table 3** for the two selected microalgae [65].

Generally, mechanical pretreatment techniques are recommended for macroalgae, while thermal techniques are the most feasible for microalgae. Among the mechanical pretreatment, the most common are chipping, milling, grinding, shredding, and cutting. The mechanical pretreatments enhance the digestibility and biodegradability of the seaweed as they can efficiently disrupt the cell wall; nevertheless, they are high-energy-wasting [66]. **Table 4** shows previous results of studies available in the literature about the effects of pretreatments on microalgae biogas production and the beneficial effect of pretreatment on biogas yields. Among them, mechanical and thermal are the most favorable ones, while microwave

Algae species Biogas yield	Operational parameters			
	Mesophilic = 35°C	Thermophilic = 55°C	pH 7.0–7.5	pH 6.5–7.0
Chlorella vulgaris	250 mL CH_4/g VS	320 mL CH_4/g VS	280 mL CH_4/g VS	260 mL CH_4/g VS
Scenedesmus obliquus	280 mL CH_4/g VS	360 mL CH_4/g VS	300 mL CH_4/g VS	280 mL CH_4/g VS

Table 2.
Effect of different operational parameters on biogas yield [15].

Algae species Biogas yield	No pretreatment	Thermal pretreatment	Mechanical pretreatment
Chlorella vulgaris	150–200 mL CH_4/g VS	350–400 mL CH_4/g VS	300–350 mL CH_4/g VS
Scenedesmus obliquus	175–225 mL CH_4/g VS	400–450 mL CH_4/g VS	350–400 mL CH_4/g VS

Table 3.
Effects of pretreatments on biogas yield of two selected algae [14].

Microalgal biomass	T (°C)	Time (days)	Biogas yield	Pretreatment	Ref.
Laminaria sp.	38	—	*615 mL g - 1 VS*	Mechanical pretreatment	[67]
Ascophyllum nodosum	38	—	*402 mL g - 1 VS*	Mechanical pretreatment	[67]
Dunaliella salina	38	32	*505 mL g - 1 VS*	Fermentative pretreatment	[68]
Botrycoccus brauni (pretreated biomass)	30	—	521 mL CH_4/g VS.	Enzymatic pretreatment	[69]
Chlorella vulgaris	36	—	369	Without pretreatment	[70]
Botryococcus braunii	35	34–50	370 ± 9 mL/ gTVSin	Without pretreatment	[52]
Chlamydomonas sp.	35	34–50	333 ± 9 mL/ gTVSin	Without pretreatment	[52]
Isochrysis sp.	35	34–50	408 ± 4 mL/ gTVSin	Without pretreatment	[52]
O-E Chlorella *minutissima microalgae*	55	30	341.43 cm^3g VS	Without pretreatment	[53]
Ulvalactuca macroalgae	55	30	342.59 cm^3/g VS	Without pretreatment	[53]

Table 4.
Biomass yield for biogas production using AD conversion process.

pretreatments have been demonstrated to have scalability issues, and the thermochemical increases the microalgae biodegradability with the risk of contaminating pilot-scale reactors [39, 71].

Algal biomass decomposition rate has been demonstrated to be increased by thermal pretreatment at temperatures between 70°C and 100°C, hence producing a 50–100% increase° in methane output relative to untreated samples [63]. Mechanical methods, like ultrasonic or high-pressure homogenization, might impair cell integrity and, therefore, release more organic substances for digestion. Co-digestion with other substrates like food waste or sewage sludge boosts the methane output by up to 30% if compared to traditional AD [10]. Furthermore, running digesters in thermophilic conditions—about 55°C—can boost microbial activity and methane production rates relative to mesophilic conditions [15]. Finally, enzymatic pretreatment can be more energy-efficient than thermal techniques, lowering energy input to 0.2–0.4 kWh per kilogram of algae while still increasing methane generation by 30–50% [72].

Diverse conversional processes for microalgal biogas generation have been techno-economically compared in **Table 5**.

Direct anaerobic digestion has a negative energy balance and produces less biogas (100–200 mL CH_4/g VS) despite cheap capital and operational expenses [63]. Although pretreated anaerobic digestion requires more capital, it can increase biogas output to 300–500 mL CH_4/g VS, improving energy balance [58, 73]. Co-digesting algae with various biomass types yields cost-effective biogas outputs of 250–450 mL CH_4/g VS and a positive energy balance due to higher digestion efficiency [74]. The

Process	Capital cost	Operational cost	Biogas yield	Energy balance	Scalability
Direct anaerobic digestion	Low	Low	100–200 mL CH_4/g VS	Negative	High
Pretreated anaerobic digestion	Medium to High	High	300–500 mL CH_4/g VS	Positive	Medium
Co-digestion	Medium	Medium	250–450 mL CH_4/g VS	Positive	High
Multi-stage digestion	High	High	400–600 mL CH_4/g VS	Positive	Low to Medium

Table 5.
Techno-economic analysis of microalgae biogas production processes [58, 63, 73].

most expensive and sophisticated method is the multi-stage digestion process that produces the highest biogas (400–600 mL CH_4/g VS), although it is not feasible for large-scale applications.

6. Perspectives and barriers that prevent from large-scale production of algae biogas

Biogas production from microalgae is still an immature process, and additional research efforts are required to overcome the existing challenges. Algae cultivation, harvesting, pretreatments, and digestion infrastructure still represent a huge cost of the process. Vertical photobioreactors and floating offshore systems are being investigated to enhance land use efficiency and sunshine exposure reducing algae cultivation costs. Studies carried out demonstrate these new technologies to be able to boost productivity up to 30% in the unit area, lowering land use and manufacturing costs [75]. Integrating algae growing with wastewater treatment or carbon capture process can reduce production costs by offering the twin advantages in terms of biogas generation and environmental restoration. Algae capacity to capture CO_2 (from 1.5 to 2.5 kg CO_2 per kilogram of dry biomass) makes them eligible to be incorporated into high-pollutant industrial processes, enhancing the economic viability of algae farming [21]. Moreover, when integrated in wastewater, algae have been demonstrated to diminish nutrient loads efficiently, attaining up to 90% removal of nitrogen and phosphorus while concurrently generating biomass for biogas and reducing farming costs [76].

A pilot project conducted in 2022 found that biogas produced by combining algae growing and municipal wastewater treatment cost 0.35 per kg of dry biomass, far less than traditional farming techniques [58, 77]. To enhance the large-scale production and profitability of algae biogas, a biorefinery approach, where multiple products are derived from algae biomass, is often proposed [78]. Biorefineries might improve algal waste value by 400% [79]. Genetic engineering is creating algae strains with higher lipid content, faster development, and environmental resilience. Moreover, recent advances in biotechnology have enabled the possibility of isolating and easily selecting specific species/strains. This genetic manipulation can be a step forward to the improvement of harvesting efficiency by improving light utilization, algae production, carbon dioxide fixation, reducing costs for cell lysis, and biomass recovery [80]. Unfortunately, just a few microalgae are known until now to be suitable for genetic manipulation. Among them, *Chlamydomonas renhardtii*, a green microalga, has demonstrated the best result, so it has been largely used for gene studies [81].

Cultivation cost	\$0.50 per kg of dry biomass
Harvesting cost	\$0.10–\$0.20 per kg of dry biomass
Energy balance (EROI)	1.2–2.5
Pretreatment energy use	0.2–0.4 kWh/kg algae
Biomass yield variability	300–350 mL CH_4/g VS
Scale-up investment	\$2 M–\$8 M for biogas facilities
Water consumption	Net zero or negative with wastewater integration
Biogas electricity price	\$0.07–\$0.12 per kWh

Table 6.
Recent data on algal biogas production barriers. Elaborated by the authors from [11–13, 22, 72, 83–86].

To be scalable, the AD process needs to reduce algae reduction of cultivation costs. The main problem is the energy required for cultivation, particularly in photobioreactors, which remains high, with costs for light and aeration often exceeding \$0.50 per kg of dry biomass [12]. Controversially, CO_2 used as a nutrient for algae can reach a minimum cost of \$0.02 per kg in industrial large-scale applications [15]. Recent advancements in harvesting technologies have reduced the operation costs of the algae dewatering, combining flocculation with sedimentation, with a cost of \$0.10–\$0.20 per kg of dry biomass [13]. A 2022 research study revealed that electro-flocculation might reduce harvesting expenses by 50% over centrifugation systems [82]. Under optimal conditions, life cycle studies revealed that the energy return on investment (EROI) for algal biogas could vary from 1.2 to 2.5, indicating a possible positive energy balance when coupled with wastewater treatment or CO_2 collection systems [83].

Moreover, it has been proved that large-scale systems can achieve high productivity (up to 30 tons per hectare per year), but they still require considerable investment. Initial investments for a medium-sized biogas plant from algae might range from \$2 million to \$8 million, but they have been demonstrated to be lowered by using modular designs [84]. Although water use is still a problem, integration with industrial effluent has proven that growing algae may lower net water use; some systems even generate extra water for reuse.

Current changes in EU regulations are supporting a circular economy, promoting biomass use, and this trend can be favorable to the generation of algal biogas [85]. Carbon pricing and green energy certificates are enabling algal biogas to become more commercially feasible. Recent projections show that biogas from algae might become competitive in the energy markets at \$0.07–\$0.12 per kWh [22]. **Table 6** summarizes the main barriers to a large-scale application of algae biogas.

As it can be seen, policy support, R&D investment, and infrastructure development will be crucial for algae biofuel delivery, as well as to address these technological and financial constraints. Technological advancements may contribute to lower prices, increased output, and the incorporation of algae cultivation into industrial processes for mutual benefit.

7. Conclusions

Algae are a neutral alternative source that can play a pivotal role in current society, leading the way toward fossil fuel decarbonization. The growing interest toward this third-generation biomass resides in its beneficial properties, mainly the high growth rates, fast development, and carbon dioxide absorption capability that give it a larger potential energy output compared to traditional fuel. Even if the algae biomass market has been predominantly dominated by Asiatic countries, nowadays also, Europe, the United States, Australia, and New Zealand have started to expand seaweed and microalgae cultivation, investing in production technologies for the large-scale production of algae biogas. The analysis carried out has shown seaweed exploitation to be more mature than the production of biogas from microalgae, which still has some limitations that need to be addressed. Pretreatment methods have demonstrated the capability to improve biomass solubility, increasing the production of lipids and proteins and, consequently, biogas yield. Moreover, the study has enhanced the potential of algae biofuel as a sustainable resource to replace traditional fossil fuels in internal combustion engines. On the other side, applications for defense are gaining special interest where, ideally, the army can cultivate algal biomass locally and use it for biofuel production. High investment and operation costs are still the main issues that need to be addressed. Policy support, R&D efforts, and infrastructure development have been identified to be crucial for algae biofuel delivery on a large scale.

Nomenclature

AD	anaerobic digestion
B100	100% pure biodiesel
C/N ratio	carbon to nitrogen ratio
EROI	energy return on investment
FAME	fatty acid methyl esters
HRT	hydraulic retention time
ICE	internal combustion engines
mL CH_4/g VS	milliliters of methane per gram of volatile solids—a measure of biogas yield.
PBR	photobioreactor
VS	volatile solids
TVS	total volatile solids

Author details

Lina Montuori*, Manuel Alcázar-Ortega and Abdelhafidh Benreguieg
Institute for Energy Engineering, Universitat Politècnica de València, Valencia, Spain

*Address all correspondence to: lmontuori@iie.upv.es

References

[1] Araújo R et al. Current status of the algae production industry in Europe: An emerging sector of the blue bioeconomy. Frontiers in Marine Science. 2021;**7**:art. no. 626389. DOI: 10.3389/fmars.2020.626389

[2] Camarena-Gómez MT et al. Macroalgae production in northern Europe: Business and government perspectives on how to regulate a novel blue bioeconomy. Aquaculture. 2022;**560**:art. no. 738434. DOI: 10.1016/j.aquaculture.2022.738434

[3] Camia A et al. Biomass Production, Supply, Uses and Flows in the European Union: First Results from An Integrated Assessment. Publications Office of the European Union, Re. no. EUR 28993 EN; 2018. DOI: 10.2760/539520

[4] Bora A et al. Microalgae to bioenergy production: Recent advances, influencing parameters, utilization of wastewater – A critical review. Science of the Total Environment. 2024;**936**:art. no. 174230. DOI: 10.1016/j.scitotenv.2024.174230

[5] IEA Bioenergy. State of Technology Review – Algae Bioenergy; 2017. DOI: 10.13140/RG.2.2.11770.90560

[6] Griffiths MJ, Harrison RPVHSTL. Selection of direct transesterification as the preferred method for assay of fatty acid content of microalgae. Lipids. 2010;**45**(11):1053-1060

[7] Razzak SA et al. Microalgae cultivation in photobioreactors: Sustainable solutions for a greener future. Green Chemical Engineering. 2024;**5**(4):418-439. DOI: 10.1016/j.gce.2023.11.001

[8] Hemavathy R, Ragini Y, Shruthi S, Ranjani S, Subhashini S, Thamarai P. Biofuel production from marine macroalgae: Pathways, technologies, and sustainable energy solutions. Industrial Crops and Products. Feb 2025;**224**:Art. no. 120282. DOI: 10.1016/j.indcrop.2024.120282

[9] Kammler S, Malvis Romero A, Burkhardt C, Barut L, Antranikian G, Liese A, et al. Macroalgae valorization for the production of polymers, chemicals, and energy. Biomass and Bioenergy. Apr 2024;**183**:Art. no. 107105. DOI: 10.1016/j.biombioe.2024.107105

[10] Bohutskyi P, Bouwer EJ. Biogas production from algae and cyanobacteria through anaerobic digestion: A review, analysis, and research needs. In: Lee JW, editor. Advanced Biofuels and Bioproducts. Vol. 2, ch. 36. New York, NY, USA: Springer; 2013. pp. 873-975. DOI: 10.1007/978-1-4614-3348-4_36

[11] Leflay H, Pandhal J, Brown S. Direct measurements of CO2 capture are essential to assess the technical and economic potential of algal-CCUS. Journal of CO2 Utilization. 2021;**52**:101657. ISSN 2212-9820. DOI: 10.1016/j.jcou.2021.101657

[12] Doshi A, Pascoe S, Coglan L, Rainey TJ. Economic and policy issues in the production of algae-based biofuels: A review. Renewable and Sustainable Energy Reviews. 2016;**64**:329-337. ISSN 1364-0321. DOI: 10.1016/j.rser.2016.06.027

[13] Shanab SMM et al. Harvesting of microalgae using natural flocculants: An update on recent developments. Algal Research. 2021;**55**:102240

[14] Passos D, García J, Ferrer I. Impact of low temperature thermal pre-treatment

on the anaerobic digestion of microalgae biomass. Bioresource Technology. 2013;**138**:285-291

[15] Mendez L, Mahdy A, Timmermans M, Fernández C. Influence of thermal pre-treatment on the physical and chemical properties of microalgae biomass. Algal Research. 2015;**12**:29-35

[16] Wahlen BD et al. Mitigation of variable seasonal productivity in algae biomass through blending and ensiling: An assessment of compositional changes in storage. Algal Research. 2019;**42**:art. no. 101584. DOI: 10.1016/j.algal.2019.101584

[17] Wahlen BD et al. Preservation of microalgae, lignocellulosic biomass blends by ensiling to enable consistent year-round feedstock supply for thermochemical conversion to biofuels. Frontiers in Bioengineering and Biotechnology. 2020;**8**:art. no. 316. DOI: 10.3389/fbioe.2020.00316

[18] Yang Y, Tang S, Chen JP. Carbon capture and utilization by algae with high concentration CO2 or bicarbonate as carbon source. Science of the Total Environment. 2024:art. no. 170325. DOI: 10.1016/j.scitotenv.2024.170325

[19] Singh J, Saxena RC. An introduction to microalgae: Diversity and significance. In: Kim S-K, editor. Handbook of Marine Microalgae: Biotechnology Advances. Amsterdam: Elsevier; 2015. pp. 11-24. DOI: 10.1016/B978-0-12-800776-1.00002-9

[20] Statistics Team (NFISS) - Fisheries and Aquaculture Organization of the United Nations, Seaweed Production; 2020. Available from: www.fai.org

[21] Razzak SA, Bahar K, Oajedul Islam KM, Haniffa AK, Faruque MO, Zakir Hossain SM, et al. Microalgae cultivation in photobioreactors: Sustainable solutions for a greener future. Green Chemical Engineering. 2024;**5**(4):418-439. ISSN 2666-9528. DOI: 10.1016/j.gce.2023.10.004

[22] Zabochnicka M et al. Algal biomass utilization toward circular economy. Life (Basel). 2022;**12**(10):art. no. 1480. DOI: 10.3390/life12101480

[23] Veerabadhran M, Gnanasekaran D, Wei J, Yang F. Anaerobic digestion of microalgal biomass for bioenergy production, removal of nutrients and microcystin: Current status. Journal of Applied Microbiology. 2021;**131**:1693

[24] Prabakaran P, Karthikeyan S, et al. Algae biofuel: A futuristic, sustainable, renewable and green fuel for IC engines. Materials Today Proceedings. 2023;**83**:172-175. DOI: 10.1016/j.matpr.2023.03.579

[25] Campbell I, Macleod A, Sahlmann C, Neves L, Funderud J, Øverland M, et al. The environmental risks associated with the development of seaweed farming in Europe – Prioritizing key knowledge gaps. Frontiers in Marine Science. 2019;**6**:107

[26] Akila V, Ayyasamy PM, Balakrishnan S, Manikandan A, Rajakumar S, Sukeetha DS. Biogas and biofertilizer production of marine macroalgae: An effective anaerobic digestion of Ulva sp. Biocatalysis and Agricultural Biotechnology. Apr 2019;**18**:Art. no. 101035. DOI: 10.1016/j.bcab.2019.101035

[27] Pardilhó S, Boaventura R, Almeida M, Dias JM. Marine macroalgae waste: A potential feedstock for biogas production. Journal of Environmental Management. Feb 2022;**304**:Art. no. 114309. DOI: 10.1016/j.jenvman.2021.114309

[28] de Carvalho JC et al. Microalgal biomass pretreatment for integrated processing into biofuels, food, and feed. Bioresource Technology. 2020;**300**:art. no. 122719. DOI: 10.1016/j.biortech.2019.122719

[29] Narala RR et al. Comparison of microalgae cultivation in Photobioreactor, open raceway pond, and a two-stage hybrid system. Frontiers in Energy Research. 2016;**4**:art. no. 29. DOI: 10.3389/fenrg.2016.00029

[30] Razzak SA, Bahar K, Islam KO, Haniffa AK, Faruque MO, Hossain SZ, et al. Microalgae cultivation in photobioreactors: Sustainable solutions for a greener future. Green Chemical Engineering. 2023;**5**:418-439

[31] Sirajunnisa AR, Surendhiran D. Algae – A quintessential and positive resource of bioethanol production: A comprehensive review. Renewable and Sustainable Energy Reviews. 2016;**66**:248-267

[32] Deprá MC, Mérida LG, Menezes CRD, Zepka LQ, Jacob-Lopes E. A new hybrid photobioreactor design for microalgae culture. Chemical Engineering Research and Design. 2019;**144**:1-10

[33] Zabed HM et al. Biogas from microalgae: Technologies, challenges and opportunities. Renewable and Sustainable Energy Reviews. 2020;**117**:art. no. 109503. DOI: 10.1016/j.rser.2019.109503

[34] Peu P, Sassi J, Girault R, Picard S, Saint-Cast P, Béline F, et al. Sulphur fate and anaerobic biodegradation potential during co-digestion of seaweed biomass (Ulva sp.) with pig slurry. Bioresource Technolonogy. 2011;**102**:10794-10802

[35] Vanegas C, Bartlett J. Green energy from marine algae: Biogas production and composition from the anaerobic digestion of Irish seaweed species. Environmental Technology. 2013;**1**:34

[36] Abusweireh RS et al. Algae biogas production focusing on operating conditions and conversion mechanisms – A review. Heliyon. 2023;**9**(7):art. no. e17757. DOI: 10.1016/j.heliyon.2023.e17757

[37] Milledge JJ, Nielsen BV, Maneein S, Harvey PJ. A brief review of anaerobic digestion of algae for bioenergy. Energies. 2019;**12**(6):1166

[38] Tiwari B, Troy D. Seaweed Sustainability: Food and Non-Food Applications. Amsterdam: Academic Press; 2015

[39] Jankowska E, Sahu AK, Oleskowicz-Popiel P. Biogas from microalgae: Review on microalgae's cultivation, harvesting and pretreatment for anaerobic digestion. Renewable and Sustainable Energy Reviews. 2017;**75**:692-709. DOI: 10.1016/j.rser.2016.11.045

[40] Sher F, Smječanin N, Hrnjić H, Karadža A, Omanović R, Šehović E, et al. Emerging technologies for biogas production: A critical review on recent progress, challenges and future perspectives. Process Safety and Environment. 2024;**188**:834-859

[41] Statista. Share of Primary Energy Consumption by Source in Spain; 2023. Available from: https://www.statista.com/statistics/

[42] Dependency on Energy Imports in Spain; 2025. Available from: https://www.statista.com

[43] Vasan V, Sridharan NV, Feroskhan M, Vaithiyanathan S, Subramanian B, Tsai P-C, et al. Biogas production and its utilization in internal combustion engines - a review. Process Safety and Environmental Protection. 2024;**186**:518-539

[44] Khayum N, Anbarasu S, Murugan S. Combined effect of fuel injecting timing and nozzle opening pressure of a biogas-biodiesel fuelled diesel engine. Fuel. 2020;**262**:art. no. 116505. DOI: 10.1016/j.fuel.2019.116505

[45] Khan O, Alsaduni I, Equbal A, Parvez M, Yadav AK. Performance and emission analysis of biodiesel blends enriched with biohydrogen and biogas in internal combustion engines. Process Safety and Environmental Protection. Jan 2024;**183**:1013-1037. DOI: 10.1016/j.psep.2024.01.049

[46] Demirbas A. Progress and recent trends in biodiesel fuels. Energy Conversion and Management. 2009;**50**:14-34

[47] Mohite A et al. Performance enhancement and emission control through adjustment of operating parameters of a biogas-biodiesel dual fuel diesel engine: An experimental and statistical study with biogas as a hydrogen carrier. International Journal of Hydrogen Energy. 2023;**52**:752. DOI: 10.1016/j.ijhydene.2023.08.201

[48] Satputaley S, Zodpe D, Deshpande N. Performance, combustion and exhaust emissions analysis of a diesel engine fuelled with algae oil and algae biodiesel. Materials Today Proceedings. 2018;**5**(11):2302-2303

[49] Sekar M, Selim MYE, Elgendi M. Improving the performance of a diesel engine using nanomaterials and chlorella vulgaris microalgae blends assisted with biogas. International Journal of Hydrogen Energy. 2024;**49**:496-506. DOI: 10.1016/j.ijhydene.2023.09.171

[50] Xu Y, Hellier P, Purton S, Baganz F, Ladommatos N. Algal biomass and diesel emulsions: An alternative approach for utilizing the energy content of microalgal biomass in diesel engines. Applied Energy. 2016;**172**:80-95

[51] Al-Ansari MM et al. Assessing the benefits of Chlorella vulgaris microalgal biodiesel for internal combustion engines: Energy and exergy analyses. Fuel. 2023;**344**(1):art. no. 128055. DOI: 10.1016/j.fuel.2023.128055

[52] Frigon J-C et al. Screening microalgae strains for their productivity in methane following anaerobic digestion. Applied Energy. 2013;**108**:100-107. DOI: 10.1016/j.apenergy.2013.02.051

[53] Koçer A, Özçimen D. Investigation of the biogas production potential from algal wastes. Waste Management & Research. Nov 2018;**36**(11):1100-1105. DOI: 10.1177/0734242X18798447

[54] Aravind S et al. Influence of compression ratio variations on hydrogen combustion dynamics in a tri-fuel agricultural diesel engine operating with diesel, algae spirogyra methyl ester, and di-tert-butyl peroxide. International Journal of Hydrogen Energy. 2025;**97**:1399-1410. DOI: 10.1016/j.ijhydene.2025.01.113

[55] Aravind S et al. Exposure the role of hydrogen with algae spirogyra biodiesel and fuel-borne additive on a diesel engine: An experimental assessment on dual fuel combustion mode. Case Studies in Thermal Engineering. 2025;**65**(8):art. no. 105566. DOI: 10.1016/j.csite.2024.105566

[56] Canakci M. The potential of restaurant waste lipids as biodiesel feedstocks. Bioresource Technology. 2007;**98**:183-190

[57] Qian Y, Sun S, Shan DJX, Lu X. Review of the state-of-the-art of biogas combustion mechanisms and applications in internal combustion

engine. Renewable and Sustainable Energy Reviews. 2017;**69**:50-58

[58] Collet P, Lardon L, Hélias A, Bricout S, Lombaert-Valot I, Perrier B. Biodiesel from microalgae – Life cycle assessment and recommendations for potential improvements. Renewable Energy. Nov 2014;**71**:525-533. DOI: 10.1016/j.renene.2014.06.009

[59] Arbib Z. Success stories of advanced biofuels for transport all-gas: Algae biofuel for vehicles. Bioenergy. 2016

[60] UWI. Barbados Launches First Bio-Cng Vehicle Powered by Aspx. Cavehill. Available from:https://www.cavehill.uwi.edu/news-events/news-releases/2024/september: UWI; 2024

[61] Ryskamp R, Carder D. The next generation of biofuels for the U.S. Military - The Homeland Defense & Security Information Analysis Center (HDIAC). HDIAC Journal. Spring. 2017;**4**(1)

[62] Bigger P, Neimark BD. Weaponizing nature: The geopolitical ecology of the US Navy's biofuel program. Political Geography. 2017;**60**:13-22. ISSN 0962-6298. DOI: 10.1016/j.polgeo.2017.03.007

[63] Passos F, Ferrer I, García J. Characterization of thermophilic anaerobic digestion of microalgae biomass. Bioresource Technology. 2014;**173**:392-399

[64] Wu N et al. Techno-economic analysis of biogas production from microalgae through anaerobic digestion. In: Anaerobic Digestion. London, UK: IntechOpen; 2019. DOI: 10.5772/intechopen.86090

[65] del Rosario Rodero M et al. Harvesting microalgal-bacterial biomass from biogas upgrading process and evaluating the impact of flocculants on their growth during repeated recycling of the spent medium. Algal Reserach. 2020;**48**:art. no. 101915. DOI: 10.1016/j.algal.2020.101915

[66] Rodriguez C, Alaswad A, Mooney J, Prescott T, Olabi AG. Pre-treatment techniques used for anaerobic digestion of algae. Fuel Processing Technology. Oct 2015;**138**:765-779. DOI: 10.1016/j.fuproc.2015.06.027

[67] Montingelli M, Benyounis K, Quilty B, Stokes J, Olabi A. Influenece of mechanical pretreatment and organi concentration of irish brown seaweed for methane production. Energy. 2016;**118**:1079-1089

[68] Mussgnug J, Klassen V, Schlüter A, Kruse O. Microalgae as substrates for fermentative biogas production in a combined biorefinery concept. Journal of Biotechnology. Oct 2010;**150**(1):51-56. DOI: 10.1016/j.jbiotec.2010.07.030

[69] Ciudad G, Rubilar O, Azócar L, Toro C, Cea M, Torres Á, et al. Performance of an enzymatic extract in Botrycoccus braunii cell wall disruption. Journal of Bioscience and Bioengineering. Jan 2014;**117**(1):75-80. DOI: 10.1016/j.jbiosc.2013.06.012

[70] Prajapati SK, Malik A, Vijay VK. Comparative evaluation of biomass production and bioenergy generation potential of chlorella spp. through anaerobic digestion. Applied Energy. 2014;**114**:790-797. DOI: 10.1016/j.apenergy.2013.08.021

[71] Passos F, Uggetti E, Carrère H, Ferrer I. Pretreatment of microalgae to improve biogas production: A review. Bioresource Technology. Nov 2014;**172**:403-412. DOI: 10.1016/j.biortech.2014.08.114

[72] Passos F, Hom-Diaz A, Blanquez P, Vicent T, Ferrer I. Improving biogas

production from microalgae by enzymatic pretreatment. Bioresource Technology. 2016;**199**:347-351. ISSN 0960-8524. DOI: 10.1016/j.biortech.2015.08.084

[73] Passos D, Ferrer I, García J, Uggetti J. Cultivation of microalgae in urban wastewater and their potential for anaerobic digestion. Chemical Technology and Biotechnology. 2014

[74] Quinn JC et al. The potentials and challenges of algae based biofuels: A review of the techno-economic, life cycle, and resource assessment modeling. Bioresource Technology. 2015;**184**:444-452. DOI: 10.1016/j.biortech.2014.10.075

[75] Kim TSJ et al. Vertical Photobioreactor Design for Enhanced Microalgal Biomass Production. Biotechnology and Bioengineering. 2021;**118**:1014-1026

[76] Demirer G, Calicioglu O. Integrated nutrient removal and biogas production by Chlorella vulgaris cultures. Journal of Renewable and Sustainable Energy. 2015;7(3):art. no. 033123. DOI: 10.1063/1.4922231

[77] Bhandari M, Kumar P, Bhatt P, Simsek H, Kumar R, Chaudhary A, et al. An integration of algae-mediated wastewater treatment and resource recovery through anaerobic digestion. Journal of Environmental Management. 2023;**342**:118159. ISSN 0301-4797. DOI: 10.1016/j.jenvman.2023.118159

[78] Razzak SA et al. Integrated CO_2 capture, wastewater treatment and biofuel production by microalgae culturing—A review. Renewable and Sustainable Energy Reviews. 2013;**27**:622-653. DOI: 10.1016/j.rser.2013.05.063

[79] Khoo CG, Dasan YK, Lam MK, Lee KT. Algae biorefinery: Review on a broad spectrum of downstream processes and products. Bioresource Technology. 2019;**292**:121964. ISSN 0960-8524. DOI: 10.1016/j.biortech.2019.121964

[80] Sydney EB et al. Microalgal strain selection for biofuel production. In: Biofuels from Algae. 2nd ed. Elsevier; 2019. pp. 51-66. DOI: 10.1016/B978-0-444-64192-2.00003-2

[81] Banerjee C, Dubey KK, Shukla P. Metabolic engineering of microalgal based biofuel production: Prospects and challenges. Frontiers in Microbiology. 2016;7:art. no. 432. DOI: 10.3389/fmicb.2016.00432

[82] Matter IA, Bui VKH, Jung M, Seo JY, Kim Y-E, Lee Y-C, et al. Flocculation harvesting techniques for microalgae: A review. Applied Sciences. 2019;**9**(15):3069, 2076-3417. Available from: https://www.mdpi.com/2076-3417/9/15/3069

[83] Bradley T, Rajaeifar MA, Kenny A, et al. Life cycle assessment of microalgae-derived biodiesel. The International Journal of Life Cycle Assessment. 2023;**28**:590-609. DOI: 10.1007/s11367-023-02140-6

[84] Chen SS et al. Scaling up of microalgae cultivation for bioenergy production: Challenges and opportunities. Bioresource Technology. 2021:335

[85] Tran KC. Anaerobic Digestion of Microalgal Biomass: Effects of Solid Concentration and Pre-Treatment. Southampton, UK: University of Southampton, Faculty of Engineering and the Environment; 2017

[86] Ward AJ, Hobbs PJ, Holliman PJ, Jones DL. Optimisation of the anaerobic digestion of agricultural resources. Bioresource Technology. Nov 2008;**99**(17):7928-7940. DOI: 10.1016/j.biortech.2008.02.044. Epub 2008 Apr 11

Chapter 4

Membrane Covers for Biogas Storage Tanks

Rosemarie Wagner and Kai Heinlein

Abstract

Membrane covers consist of foils and coated fabrics for the storage of biogas. These are mounted above pre-fermentation tanks, fermenters, and secondary fermentation tanks. The covers are used *via* digestate storage to prevent the spread of fermentation odour into the environment. They are economical, which can be installed in a short time and react more flexibly when producing, storing, and extracting biogas. The covers have been in use for about 30 years, but a lack of experience in planning, manufacturing, mounting and operation has led to premature failure over the last few years. Own investigations of damage and studies on the interaction of external influences and biogas state on a test tank will be presented with the intention of reducing and avoiding defects and damage to the membrane covers over a reasonable service life. Depending on the design, different polymer materials are used. The properties have to meet the requirements of the environment, the aggressive climate in the tanks and the operation. Knowledge of the interaction between environmental influences and operating conditions reduces defects and damage. This is presented on a test tank on the basis of measured quantities, as examples of critical situations. The tank was equipped with the typical devices used in biogas tanks, measuring devices, wind station and PC with enough capacity to store the measured data.

Keywords: biogas storage, biogas tanks, membrane cover, foils, coated fabric, shapes, material properties, operation behaviour

1. Introduction

Membrane covers are used on biogas plants *via* pre-fermentation tanks, fermenters, secondary fermentation tanks and digestate storage facilities. These are low-pressure systems with an overpressure inside the covers of 200 Pa to 500 Pa, as experienced value of practise in Germany; in exceptional cases, the maximum internal pressure is 1000 Pa, depending on the size, storage volume and number of layers. At an atmospheric pressure of 1013 hPa in relation to sea level, this corresponds to 0.2–0.5% of atmospheric pressure. In order to avoid flaring in case of overproduction of biogas, the biogas is also collected in external membrane storage systems. The advantages of membrane cover compared to other designs are their low weight, flexible storage volume, cost-effective production, and time-saving mounting. The flexibility and folding allow it to be adapted to different fill levels. Common materials are made of polymers that are resistant to acidic and basic environments.

For covers over tanks, both mechanically prestressed high-point systems and internal pressure-supported (pneumatic) systems are used. In the last 30 years of the development of biogas plants, both construction methods have prevailed against rigid covers made of steel, reinforced concrete or fibre-reinforced plastics for the storage of biogas. In the early years, the planning, production and installation of the covers were carried out using simple methods. The knowledge and experience from membrane architecture were missing. The covers are over circular tank cones or spherical segments, **Figure 1**.

Films and coated fabrics were used by other sectors of the economy with a much less aggressive environment, such as tent halls or truck tarpaulins. The requirements for durability, long-term service life and gas tightness are considerably higher for membrane covers on biogas plants. The stresses in the membranes are to be calculated using simple equations. Cones allow for easy production because they can be unrolled without distortion into the plane. Spherical segments are made of equal strips in the shape of orange slices. Usually, one strip is covered with triangles on the spherical surface. Unrolling the triangles one after the other results in the cut of the strip [1]. In the past, the consequences were defects and damage to the covers. In order to make planning, production and operation safer, a recommendation on biogas membrane storage systems was prepared in Germany and published in 2016 [2]. This incorporates experience and knowledge from membrane architecture.

Membranes only carry tensile stresses, wrinkles occur under pressure or shear stress, and the membranes lose their ability to safely transfer loads. The geometry is established as an equilibrium between two normally related tensile stresses on the surface. This internal stress condition is achieved by prestressing the membranes or applying internal pressure. Without this internal state of stress, great deformations result under the influences, up to constant movement and fluttering in the wind. The surfaces are flat, uniaxial or biaxially curved. Flat or uniaxially curved surfaces tend to deform larger under loads. A biaxial curvature requires the least preload or

Figure 1.
Typical membrane covers at biogas plants [© Rosemarie Wagner].

preload to be sufficiently load-bearing. The application of the internal stress state already leads to elastic strains during construction: both the internal stress state and the geometry change. To avoid this, the cutting strips are cut out of flat foil or coated fabrics slightly smaller and are referred to as compensation. This is determined in two-axis tensile tests. The intersecting threads result in an interaction between both directions in the membrane. Depending on the type of film or fabric, the strains differ significantly from the strains in the uniaxial tensile test. If this knowledge, developed in the second half of the twentieth century for membrane buildings, is missing [1], defects and damage to the covers are the result.

2. Membrane storage systems

Covers over tanks and external membrane storage systems are presented. Depending on the function of a biogas plant, the covers consist of one, two or three independent layers.

2.1 Single-layer membrane covers

The requirements for the membranes are higher in single-layer covers than in two-layer ones. The membrane must be able to withstand all external influences over the long term, and it must be sufficiently methane-dense for service life. Damage to the membrane immediately results in a loss of biogas and reduces the load-bearing capacity.

Single-layer membranes are either cone-shaped systems that span between a mast in the middle of the tank and the upper edge of a circular tank or internal pressure-supported cone shapes and sphere segments. If mechanically pretensioned covers are based on a cone geometry, large elongation occurs in the membranes under external loads and in case of overproduction of biogas, especially in the circumferential direction. The consequence is an irreversible strain in the membrane. This leads to increased fluttering as a result of gusts of wind up to bags of water and snow, **Figure 2**. The reason is the lack of pretension in the circumferential direction for uniaxially curved cone shapes. After anchoring to the edge of the tank, the membranes are pretensioned by extending or raising the central mast. The transverse elongation in the membrane leads to strain in the circumferential direction when the membrane is tensioned.

If a geometry corresponds to a rotationally symmetrical hyperboloid instead of a conical geometry with uniaxial curvature, it has a biaxial or counter-directional

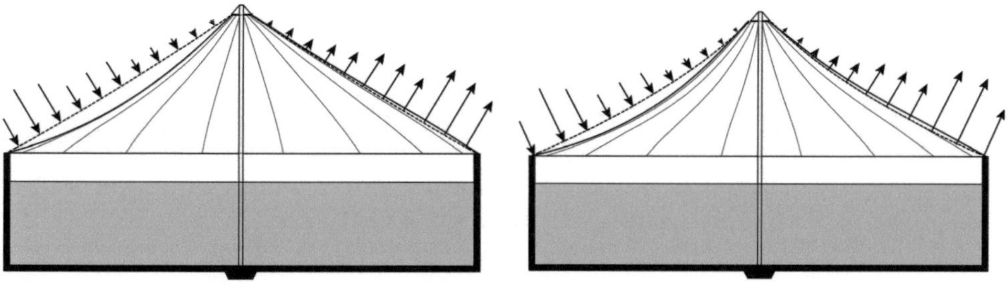

Figure 2.
Deformation with uniaxial (left) and biaxial (right) pretension under wind load [© Rosemarie Wagner].

curvature. During tensioning, additional preload is introduced into the circumferential direction and increases the stiffness. During gas production, high internal pressure leads to breakthrough and overstretching of the membrane in circumferential direction. Another disadvantage is the concentration of pretension and external loads at the top of the mast. This highly stressed area requires special design of the mast head, such as reinforcing the membrane with a second layer and avoiding too narrow or overlapping seam spacing of the individual cutting strips. A sufficiently dimensioned foundation is required in the middle of the tank under the mast. These high-point-supported surfaces are also sensitive to geometry inaccuracies. If the edge of the tank and the mast position are not checked before the membrane is manufactured, and if it is cut to an exact geometry and welded together, an inhomogeneous stress state results, which affects the durability. One advantage of this type of cover is that it maintains stability, even if the tank is empty or no biogas is produced. These are therefore preferably used to cover digestate stores.

Internal pressure-supported membranes only receive their load-bearing capacity if the internal pressure is higher than its weight or if external influences such as wind and snow. In addition, for all internal pressure-supported structures made of membranes, there is a direct correlation between the internal pressure, the geometry and the volume, the strengths of the membrane materials and load carrying. These pneumatic structures tend to have simple geometric volumes such as cylinder, cone, ellipsoid, and sphere. Cones, **Figure 3** and spherical segments are the most used covers above round tanks. The volume of a cone is about 25% smaller than a spherical segment with the same height and over the same tank. The advantage of cone-shaped covers with radial cutting patterns is lower stresses at the cone tips compared to the pole in spherical segments.

The internal pressure is the decisive factor for the stability of the covers. However, this is not a constant quantity; it changes not only with the gas level in the tank but with the temperature in it and in the environment, the solar radiation, the surface temperature of the membranes, the radiation of heat through the moist air in the tank

Figure 3.
Cone-shaped membrane as internal pressure-supported structure [© Rosemarie Wagner].

and operating conditions. This requires control of the internal pressure in order to maintain the load-bearing capacity permanently.

If the internal pressure is less than 15% of external impacts such as wind gusts or snow load, very large deformations occur until the membrane cracks or collapses. Too high internal pressure requires the less environmentally friendly flaring of biogas or leads to overstrain of the membrane. As soon as there is no internal pressure, the membrane must be secured above the substrate of a construction. This is either made of wood or consists of a mast standing in the middle of the tank. Belts are attached to this that stretch radially to the edge of the tank. Recently, a close-meshed web has been laid on the belt system.

Approximately, there is a simple relationship that the stress in a spherical segment is proportional to the internal pressure Δp and the radius of curvature R. In a flat, spherical segment, the radius of curvature R_1 is greater than in a high one with R_2. At the same internal pressure $\Delta p_1 = \Delta p_2$, the membrane stress in the flat, spherical segment is greater by the ratio of the radii of curvature, **Figure 4**.

The relation between radius R, diameter D, height H and internal pressure *Äp* is given by:

$$R = \left(\frac{D^2}{8H} + \frac{H}{2} \right) \tag{1}$$

The radius increases according to the reduction of height.

Without material properties and cutting pattern, the stress in the membrane in meridian direction σ_m and circumferential direction σ_r are the following:

$$\sigma_{m,r} = \Delta p \cdot \frac{R}{2} \theta \tag{2}$$

With the same diameter and internal pressure in **Figure 4**, the stress ratio of both geometries is in the meridian direction:

$$\sigma_{m,1} = \sigma_{m,2} \cdot \frac{R_1}{R_2} \tag{3}$$

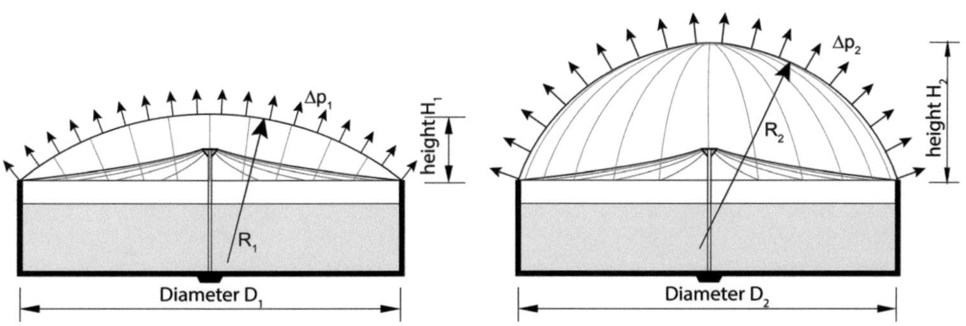

Figure 4.
Relation between curvature R and internal pressure Δp [© Rosemarie Wagner].

The disadvantage of high covers compared to low ones is the more sensitivity against wind gusts. They react with larger deformation and changes of internals pressure.

For spherical segments with parallel cutting pattern, flat covers with a ratio of height H to diameter D between 1/8 < H/D < 1/5 are suitable. Studies have been made to apply this cutting pattern to high coverage [3]. The advantage of parallel cutting is to avoid stress concentrations in the centre of the cover. This is designed as a circular polar cap. The cutting strips are welded radially to it. The seam spacing shall be greater than five times the seam width between two adjacent strips.

The disadvantage of single-layer covers is that the internal pressure is permanently maintained and requires continuous gas production. The advantage is that fill levels and thus the internal pressure can be seen directly on the geometry of the cover.

2.2 Double-layer membrane covers

Two-layer covers consist of a weather-resistant outer membrane with a resistance against all impacts. Below this is the air support volume, which is bounded downwards by the gas membrane. The internal pressure is maintained by a blower. The gas membrane separates the support air space from the gas space. If a foil is used for the gas membrane that has a significantly lower strength than the outer membrane, it is made larger to fit the outer membrane. Both mechanically pretensioned two-layer membranes and spherical segments are built. The storage volume is significantly higher in spherical segments. The disadvantage is the installation of the support structure in the absence of internal pressure due to biogas, **Figure 5**.

2.3 Three-layer membrane covers

Depending on the substrate, bacterial cultures and process, the temperatures in fermentation tanks are between 25°C and 58°C [4]. Depending on the location and function, free-standing tanks are heated and insulated. Due to the thickness of 1 mm to 2 mm of the membranes, the heat transport through the membrane is very high. With a temperature gradient of 1°C = 1 K between the inside and outside of the membrane and an average thermal conductivity of $\lambda = 0.3$ W/(mK), the heat flux for a 1 mm membrane is 300 W/m^2 and for a 2 mm membrane is 150 M/m^2. This applies on hot days, heat from the environment into the storage and on cold days and nights from the storage system in the environment. In addition, there is the heat from solar

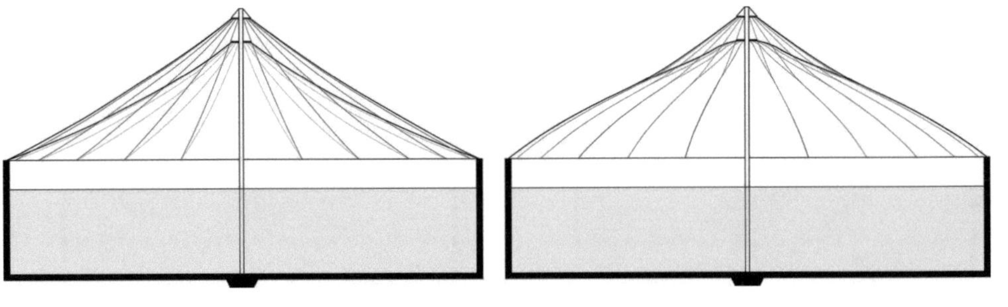

Figure 5.
Two-layer mechanically pretensioned cover, without gas (left) and with gas (right) [© Rosemarie Wagner].

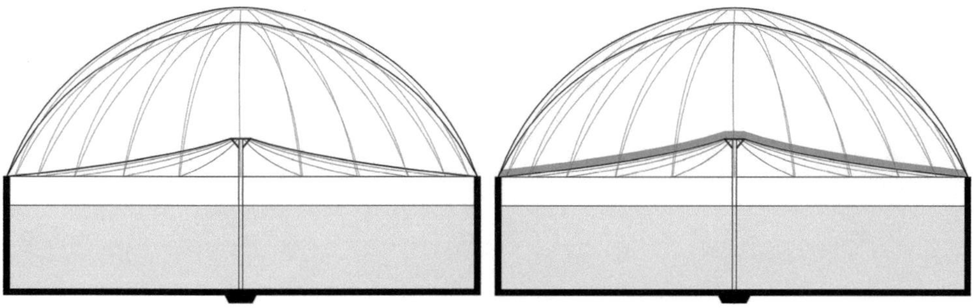

Figure 6.
Double layer membrane without and with a third layer onto the belt system as thermal isolation [© Rosemarie Wagner].

radiation and the cooling from the radiation of heat into the clear, cloudless sky [5]. In order to reduce high heat transport in and out of the fermentation tank, a membrane is attached gas-tight to the belt layer, on which insulation is placed. Biogas is fed from the fermentation into the gas chamber and stored. The requirements for insulation are very high because the insulation must not show any loss of insulation properties even at 100% moisture, the condensed water must be collected and drained away, and the insulation must be resistant to the constituents of the biogas, **Figure 6**.

3. Membrane materials

There are very few building materials that are suitable for the production and storage of biogas in the long term. These include films with a high resistance to acids and bases made of polyethylene low density (PE-LD) with a thickness of approx. 0.2 to 0.8 mm or ethylene propylene diene rubbers (EPDM) with a thickness of 2 mm. For coated fabrics, PVC is preferred as coating and polyester as fabric. The properties of membrane materials are very diverse and different for long-term use.

The information available on data sheets of producers are weight, tear strength, elongation at break, resistance to tear propagation, methane tightness at 23°C and without tension, fire resistance, temperature stability and surface resistivity at 23°C/30% rel. Humidity. The peeling and shear strength of welded or glued seams is mentioned in the data sheets for films. The adhesion is given between coating and fabric for welding seams of coated fabric. For example, PVC, as a polar material, can be welded by high frequency, thus achieving a seam strength in a uniaxial tensile test that is approximately equivalent to the tensile strength of the membrane itself. The disadvantage of the high-energy surface is a static charge. This means a risk to biogas plants because methane is flammable. The high surface energy leads to the deposit of dirt without an additional coating. Fungi and lichens grow on the weathered outside. This changes the radiation properties. Equipping it with a dirt-repellent coating and antistatic finish decreases the effort in welding. Both PE-LD and PVC have an oxygen permeability of between 800 and 1000 $cm^3/(m^2{\cdot}bar{\cdot}d)$ and 23°C [6]. The gas chamber contains almost no oxygen and the result is oxygen permeation in it. Oxygen reacts with the hydrogen sulphide to sulphur and leads to sulphur deposits on the inside of the gas membrane, **Figure 7**.

Figure 7.
The sulphur is deposited on the insulation in a three-layer membrane with condensation (left) and on the inner side of the gas membrane (right) after one year in use [© Rosemarie Wagner].

3.1 Thermoplastic films made of PE-LD

Polyethylene low density (PE-LD) is a thermoplastic material consisting of long hydrocarbon chains with amorphous and semi-crystalline arrangements. The arrangement of the crystalline areas is without orientation. Under tension, the amorphously connected molecular chains are pulled apart. This leads to an alignment of the crystalline areas in the direction of tension and to large strain. The characteristic stress-strain behaviour shows little strain at low stresses. Depending on the material, alignment occurs between 30% and 50% of the tensile strength with a significant increase in elongation and load increases A yield strength can be seen, but it runs over a range of elongation and explains the trilinear behaviour. Depending on the thickness of the film, the tensile strength is between 12 and 20 MPa, and the elongation at break is between 600% and 2000% depending on the thickness and production. The stress-strain behaviour of a PE-LD film for biogas covers is shown as an example with a thickness of 0.8 mm, carried out at the inhouse uniaxial testing machine. The results are derived from uniaxial tension tests according to DIN EN ISO 527 [7] on strips of 50 mm x 10 mm and a loading speed of 50 mm/min. The tested films had been cut out of a sample of 1 m x 1 m. Tensioning the films in the direction of production (MD) leads to a less alignment of the molecular chains. The behaviour is slightly anisotropic, with a higher stiffness in production direction (MD) and blue curves, compared to the behaviour perpendicular to the direction of production (TD) and red curves, **Figure 8** left. The thickness varies with a derivation of +/– 3%. This leads to a dispersion in behaviour, tensile strength and elongation at break. **Figure 8** shows two test results in both directions, which were taken from the sample at different positions.

PE-LD is a non-polar polymer with a low-energy surface [8]. This can be seen by water drops on the surface that roll off. This positive property has the disadvantage that the welding of two film strips is associated with effort using heat welding. Heating during welding changes the strength and stress-strain behaviour in the seam area. The tested film with seam becomes more brittle and stiffer, as shown in **Figure 8** left with the green curves. Stresses up to approx. 7.5 MPa is assumed as the severability state for the tested film. After the first loading and unloading, there is a permanent elongation of approx. 2% MD and 2,5% TD. With increasing load cycles, the remaining elongation increases slightly, **Figure 8** right. This behaviour must be considered when making the covers from foils. If the stresses are higher than the yield strength,

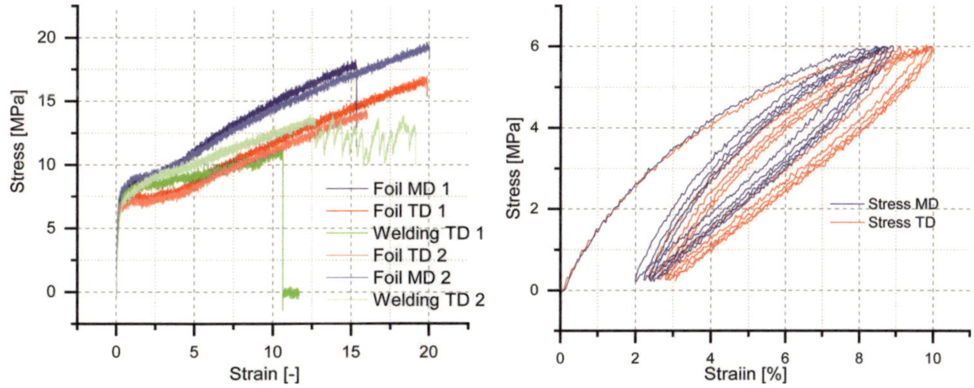

Figure 8.
Strain-stress-behaviour of a 0,8 mm PE-foil, loading till tensile strength is reached (left), load cycling (right) [© Rosemarie Wagner].

depending on the dimensions and the initial load, the permanent strains lead to larger surfaces after unloading.

Uniaxial tensile tests provide an approximation to model the behaviour under external load and internal pressure. Depending on the geometry and load, a biaxial stress state is created in the foils. If the change in the film thickness is neglected under load, different behaviours can be seen depending on the tension ratio in the production direction and perpendicular up to the yield stress. The Poisson ratio leads to the fact that the strains are determined by the biaxial stress state. **Figure 9** shows the influence of the stress ratio on the strain in both directions. The test was carried out by the in-house biaxial testing device on a cross-shaped test specimen of 320 mm x 320 mm, arm length of 120 mm and test field of 80 mm x 80 mm. The strain was measured by an optical device. The loading speed was 50 mm/min. The foil was the same as tested in **Figure 8**. The diagrams are derived from the stress and strain behaviour given in **Figure 8** up to 3,75 MPa. The result leads to a three-dimensional diagram. The horizontal axis describes the stresses in MD and TD. The vertical axis shows the strain, left MD and right TD. The stress ratio had been MD:TD = 1:2; 1:1,5;1:1, 1,5:1, 2:1 according to DIN EN 17117 [9]. The surfaces are extracted from these curves.

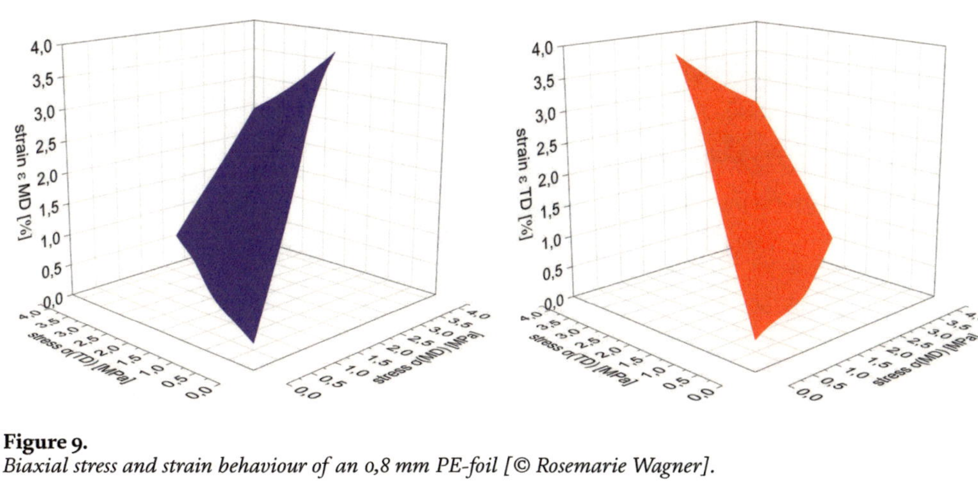

Figure 9.
Biaxial stress and strain behaviour of an 0,8 mm PE-foil [© Rosemarie Wagner].

For a stress ratio of MD:TD with 2:1 or vice versa, the strain becomes larger in the MD direction and smaller in the TD direction. The influence of the Poisson ratio is shown in the comparison of the stresses to a uniaxial tensile test. At the same tensile stress, the strains for a uniaxial tensile test are greater by a factor of 1.5 to 2 compared to a biaxial stress ratio of MD:TD with 2:1 or MD:TD with 1:2.

The amorphous content in PE-LD films leads to a sensitivity to temperature effects. The tensile strength decreases by a third when the temperature rises from 20–40°C [10]; accordingly, the strain increases. The biaxial strain has an impact on methane permeability. This increases threefold in PE-LD covers for an initial temperature of 23°C and a temperature increase of 18°C [11]. The ageing of PE-LD films is mainly caused by the action of oxygen, temperature over 40°C and UV radiation. This leads to embrittlement and a decrease in tensile strength and elongation. Direct exposure to the weather requires additives that prevent ageing. PE-LD films have a very low electrical conductivity. They are used as an inner membrane in two-layer covers or for the third layer also as PE foam and film on a belt layer.

3.2 Elastomer-foil

In films made of EPDM as an elastomer, the hydrocarbon chains are cross-linked with each other. The production is carried out in the calendar process, and these are thicker than PE films. The tensile strength and elongation at break are slightly lower than PE-LD films. Cross-linking by a chemical bond means that EPDM is one of the materials with hyperplastic behaviour. Very large elongations of up to 500% are reversible and lead to only very small permanent elongations. The trilinear stress-strain behaviour comparable to PE-LD films is evident during loading and unloading. The high ductility is suitable for biogas storage facilities because the volume adapts to gas production and extraction through reversible elongation.

To prevent overstrain, covers made of EPDM films should have a plastic mesh on the outside. Alternatively, the fill level is monitored by cable lying on the film with weight, **Figure 10**. EPDM has as positive properties such as better heat resistance,

Figure 10.
Single-layer membrane cover made of EPDM [© Rosemarie Wagner].

weather and ozone resistance compared to PE-LD films. Fillers such as carbon black and plasticisers such as mineral oils make EPDM more flexible and weather-resistant, and its flammability increases. As a non-polar material, it is water-repellent and has a low electrical conductivity. The production of seams is done by vulcanisation [12].

EPMD films also show a dependence of methane emissions on temperature. At an initial temperature of 23°C, the permeability of EPDM films increases threefold when the temperature rises to 41°C. Gas permeability is reduced by additional fillers in film production [13]. Other influences on methane emissions are the foil thickness and the strains. With an elongation of 40%, the permeability increases by about 25%. Weathering of the film over 2.5 years does not lead to a further increase in the emission rate [14]. EPDM films are preferably installed in single-layer covers.

3.3 Coated fabrics

Polyvinyl chloride (PVC) as a coating is the third most produced plastic after PE-LD and polypropylene (PP) [15]. The load carrier fabric, made of polyester (PET), occupies fifth place in global plastics production. Both PVC and polyester have no flame resistance without additives. Heat, moisture, oxygen and UV radiation lead to premature ageing. For both raw materials, additives such as boron and phosphates are used to achieve flame retardancy. Additives to reduce the ageing process are soot, titanium dioxide or sulphur-containing compounds [8]. Ageing due to auto-, thermo- and photooxidative processes cause embrittlement, reduction of tear strength and elongation at break. Both PVC and PET are resistant to weak alkalis and weak acids. PET is more susceptible to alkalis than PVC [16]. Cost-effectiveness in production and processing as well as resistance to weak acids and bases are the reasons for using PVC-coated PET fabrics. PVC is formed from rock salt and petroleum. Caustic soda and hydrogen are produced during production. It is brittle up to a temperature of 0°C [17]. In order for PVC to be turned into films and coatings on textile fabrics, plasticisers such as phthalates or epoxied vegetable oil are necessary [8]. In addition, inert and solid fillers such as calcium carbonate ($CaCO_3$) are used to increase methane tightness. Errors in the coating process lead to pre-damage. These are often hardly recognisable. In **Figure 11**, small pieces of lime can be found in the welded seam. These have an unfavourable effect on the seam strength because there is no adhesion between the coatings of the attached strips.

Additives are integrated that PVC becomes soft between 60°C and 70°C. This results in the need to test welded seams for strength even at a temperature of 70°C.

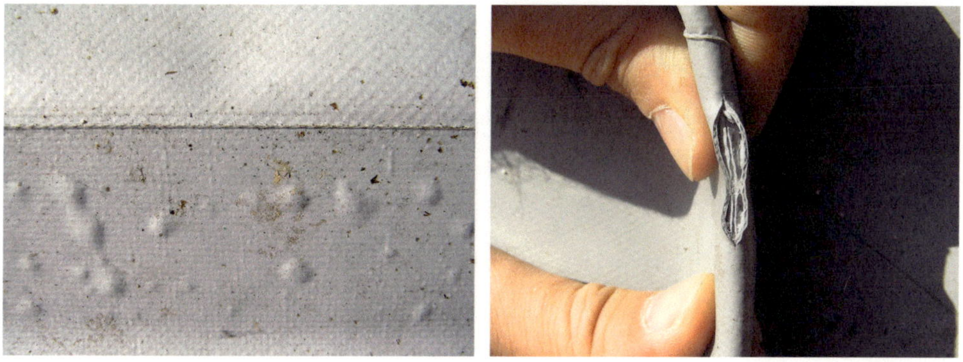

Figure 11.
Seam with pieces of lime in the coating [© Rosemarie Wagner].

The decrease in tensile strength is approximately 25–50% of the membrane itself. The dispersion depends on the seam width, the welding process and the top coat. In Germany, light grey membranes are recommended to be used for biogas covers [18]. These must be questioned with regard to their integration into the landscape, especially since the absorption of solar radiation increases with the degree of surface roughness and pollution. PVC has high absorption in the near and far infrared range, regardless of the colour [19]. The heating of the membrane is a superposition of solar radiation from the outside and thermal radiation from the inside. For example, on hot summer days, temperatures of up to 75°C were measured on a green membrane [20].

Depending on the surface quality of the PVC against the atmosphere, the plasticiser begins to migrate, and the surface silts up and becomes embrittle, **Figure 12**. The consequences are the adhesion of dirt and the appearance of cracks. The side facing the weather is provided with a varnish or protective film for permanent use. Among other things, polyvinylidene fluoride (PVDF), also polar, is used as a protective film. This allows HF welding. The PVC bottom is welded to the PVDF top. The protective layer differs in gloss from uncoated PVC.

Studies on CH_4 emissions from PVC-coated fabrics show that methane occurs mainly at fastenings to the edge of the container, level measurement and connections to rigid components [21]. In recent years, measurements have been carried out on membrane storage systems, but there is no attribution of the leakage rates of methane to the number of membranes, the membrane material, maximum temperatures, maximum stresses and service life of the membrane, as these have been done for PE-LD films and EPDM films [22]. Methane emissions are prevented by the coating. If this is damaged or destroyed down to the polyester fabric, it increases.

Polyester fabric is available on the market for membrane covers with uniaxial tensile strengths in warp/weft direction between 65 kN/m/60 kN/m and 200 kN/m/180 kN/m. The uniaxial tear strengths of EPDM films are between 11 kN/m and 15 kN/m. PE films achieve a tear strength of about 14 kN/m. The high tensile strength of the fabric is determined by membrane strips that are carried out after the coating has been applied or before assembly. At the end of the twentieth century, there were studies on the hydrolysis sensitivity of polyester fabrics [23, 24]. Long molecular chains of polyester are decomposed when exposed to moisture, amplified by higher temperatures in a weak, acidic or alkaline environment. The result is a decrease in tensile strength and embrittlement. This process is recognisable by

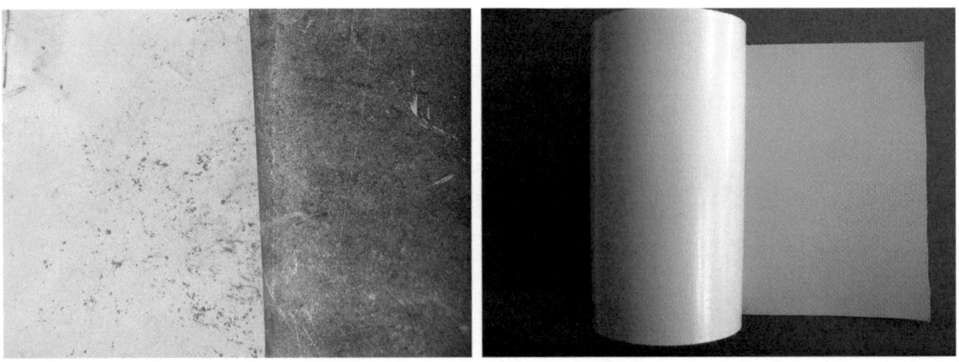

Figure 12.
Inside and outside of an approx. 5-year-old cover (left), PVC coating with PVDF layer on one side (right) [© Rosemarie Wagner].

the yellow-brownish discoloration of the yarns. Damage to the coating, especially in the case of gas membranes, leads to a reduction in the tensile strength of the fabric. As shown in **Figure 13**, the stress-strain behaviour of a membrane cover failed after 5 years. The membrane was connected to the ground. The tests were carried out with the in-house uniaxial testing device according to DIN EN DIN EN ISO 1421 [25]. The wick effect of polyester attracts moisture from the soil, so the transition from gravel to membrane tends to be very hydrolysed, especially on the sides exposed to solar radiation. **Figure 13** shows this at one of the external gas storage facilities anchored in the ground. The tensile strength decreased by 66% on the southwest and west sides with a shallow position of the sun and directly on the road. Elongation at break is reduced by 46%. This behaviour is independent of the colour of the membrane and also occurs with a light grey coating.

3.4 Biaxial tension-strain behaviour

The interweaving of warp and weft threads leads to an interaction which differs depending on the tension stresses. The behaviour is also related to whether the fabric is tensioned in the weft direction during coating.

In both single-layer and multi-layer covers, depending on a double curvature, stresses arise in the warp and weft direction as a result of pretension, internal pressure and external impact due to wind, snow and temperature. An unloaded, coated fabric behaves differently than one that has already been loaded several times. After three to five load cycles, there is permanent elongation. This changes the geometry of the membrane. If the membrane is manufactured on a sphere. This leads to a change in geometry under internal pressure. The strain across the circumferential direction increases, and the membrane becomes flatter in the polar region, **Figure 14**. The low curvature in the pole leads to high stresses, especially in the seam at the transition between radial cutting paths and the polar cap. The result is an early failure in superposition with other influences such as high internal pressure, wind and temperature.

Figure 13.
Strain-stress-behaviour depending on the orientation of the membrane to the geographical direction [© Rosemarie Wagner].

Figure 14.
Elliptical shape caused by the stresses in the membrane [© Rosemarie Wagner].

The determination of the stress–strain behaviour in biaxial tensile tests is standardised in Germany [9]. For this purpose, a load history is first defined from the maximum and minimum stresses in the warp and weft directions as a result of a structural analysis. The proof of biaxial stress–strain behaviour and serviceability is based on mean values from the different load cases and load case combinations. These are individual for each cover, depending on geometry, location and applied loads. Then, a cross-specimen is clamped in a biaxial test device and loaded according to the minimum and maximum stresses in the serviceability state, given as a force-time diagram, **Figure 15**.

Depending on the structure and production of the coated fabric, the stress-strain behaviour differs in the warp and weft direction. The diagrams, in **Figure 16**, show a

Figure 15.
Biaxial testing equipment, unloaded (left) and loaded (right) [© Rosemarie Wagner].

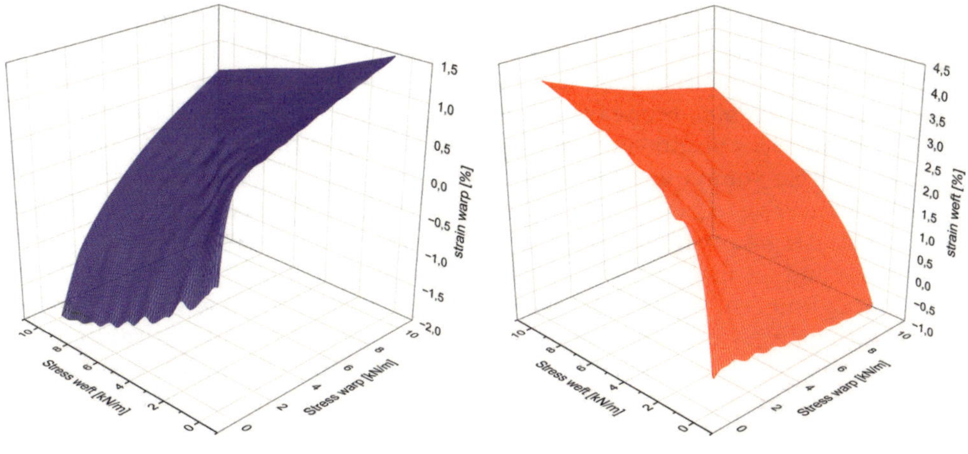

Figure 16.
Biaxial stress and strain behaviour, strain in warp direction (left) and weft direction (right) in relation to the stresses in warp and weft [© Rosemarie Wagner].

non-linear behaviour of the tested fabric with the in-house biaxial testing device. The test was carried out for a membrane applied to circular, medium-sized biogas storage tanks of app. 25 m in diameter. The total size of the specimen is 850 mm x 850 mm, the biaxial loaded area is 600 mm x 600 mm, and the field of optical measuring is 120 mm x 120 mm with 9 points of strain tracking. Depending on the pretension in warp and weft direction, the strain differs, **Figure 16**. The maximum load was in both directions 10 kN/m. The pretension was reduced from 5 kN/m to zero in steps of 1,25 kN/m. At each step, the stress in weft direction war kept constant, in warp direction, the stress was increased from the pretension level to the maximum load of 10 kN/m and vice versa.

The amount of the pretension or the internal pressure and the stresses in both directions have a high influence on the strain behaviour. The higher the pretension, the lower the elongation at the same maximum load. This means that when the biogas storage tank is full, the covers react with lower deformations but higher stresses under external impact. This can also lead to overloading and cracks in the membrane if unfavourable weather and operating conditions happen at the same time as full gas storage.

4. Operation situations

The disadvantage of two- or three-layer membrane covers is that information about the condition of the gas membrane can only be obtained *via* a small window. An indication of defects in this is the methane emission over a longer period of time at the air flaps for the air support space and along the fastening at the edge of the tank. The reason for the increase in leakage can only be determined if the membrane is dismounted and examined.

On a double-layer test tank, the behaviour of the gas membrane was investigated by simulating the five different operating states [26]. The test tank had a diameter of 15 m, a height of the outer membrane of 7,4 m and gas membranes of 6,9 m. The covers had been of biogas membrane material of a tensile strength warp/weft 68 kN/m/ 60 kN/m. The tank was equipped with a ventilator, and overpressure valves were used in practice. Measuring devices for temperature, internal pressure and volume flow in and out of the chambers had been integrated and connected to a PC with enough storage capacity. For safety reasons, the gas volume was also operated with air and without a fermentation substrate. The gas room was accessible. This was used to observe the movement of the gas membrane from one operating state to another, **Figure 17**.

With regard to gas production, an empty gas volume without substrate and biogas as well as a full gas volume and a completed gas production or no further increase in internal pressure due to gas production, can be assumed to be stationary. The addition of substrate, gas production and extraction are transient states. Depending on the substrate and gas masses, the volumes in gas and air support space change. The permanent tightness of the gas space is necessary to prevent methane from escaping into the atmosphere. Consequently, the internal pressure and temperature also change with the volume according to the thermodynamic state changes of ideal gases. Empty and filled gas space without gas production corresponds to a closed volume. The internal pressure changes over time at a constant volume when solar radiation on the outer membrane leads to the heating up of the membrane and the air support space. In order to maintain the internal pressure, volume flows in and out of the support air space must be adapted to the changing variables.

Filling with substrate, gas production, and extraction are open systems in which the substrate, gas and air are to be considered as volume flows. The thermodynamic state changes are to be extended by the time-dependent changes in the operating conditions.

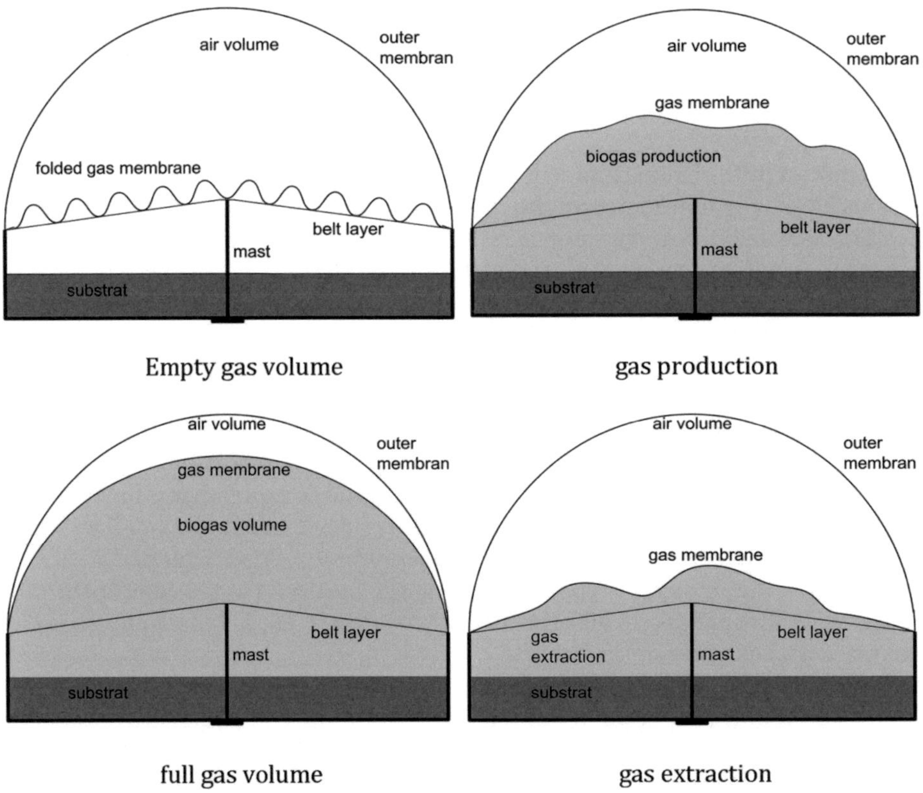

Figure 17.
Operation states of the gas membrane [© Rosemarie Wagner].

The mathematical verifications of the stability of textile covers are carried out on the basis of mechanics and structural analysis without considering variables that vary over time. Consequently, these verifications may only be provided for one empty and one full gas container without gas production. There are no reference values for the superposition with unfavourable operating conditions, especially since gas production and gas extraction are different for each plant [27]. The design of the supply air blower for pressure control in the support air space, the exhaust air openings, the overpressure and under pressure protections, the volume flows of the substrate and volume flows of the gas depending on ambient conditions is carried out according to different approaches. Assumptions for the time-dependent supply air and exhaust air volume into the support air space are based on the empirical values of the companies that offer biogas plants on the market. The same applies to gas production and gas withdrawal. Based on this experience, the blowers, pumps, diameters for the supply and exhaust air openings, the openings and pipes for the substrate supply and the gas withdrawals are designed.

In the following, different operating states are shown, and their influences on special features with regard to the gas membrane are pointed out and derived from the test tank mentioned above.

4.1 Empty gas volume

The gas volume is empty for maintenance work on the technical equipment, on the reinforced concrete or steel tanks or on the support structure. The gas membrane lies on the support structure and is stressed by the internal pressure in the air support space of a double-layer cover. If this is also without internal pressure, the outer membrane is also on top of the gas membrane. Depending on the base of the tank and the surface of the gas membrane, it folds onto the supporting structure. The formation of wrinkles results from the ratio of the surface area when filled to the base of the tank. With a constant diameter, this increases quadratically with the height of the spherical segment. For a hemisphere, the surface area is twice as large as the base area. When depositing, folds mainly occur in the circumferential direction, starting at the edge of the tank and increasing towards the middle until the entire membrane rests on the support structure. The wrinkling is imprinted on the coated fabric because the pleat pattern remains almost the same without any further influences after several loading and unloading, **Figure 18**.

The tangential wrinkling leads to kinks, and the reason lies in the change in the length of circumference. The membrane also compresses the folds in the circumferential direction. The resulting kinks lead to a local weakening of the coating over time.

The imprinting of the folds and creases is reinforced by an existing internal pressure in the air support space. In order to avoid overload due to internal pressure in the air support space and negative pressure in the gas space on the support structure and folded gas membrane, this situation requires adjusting the volume flow in the air support and gas space. Reducing the internal pressure in the air support space is only possible at low wind speeds. The low internal pressure leads to flutter at higher wind speeds.

4.2 Filling with substrate

When filling with substrate, the air in the gas chamber must escape because the gas chamber has a high methane and CO_2 tightness. The consequence is an

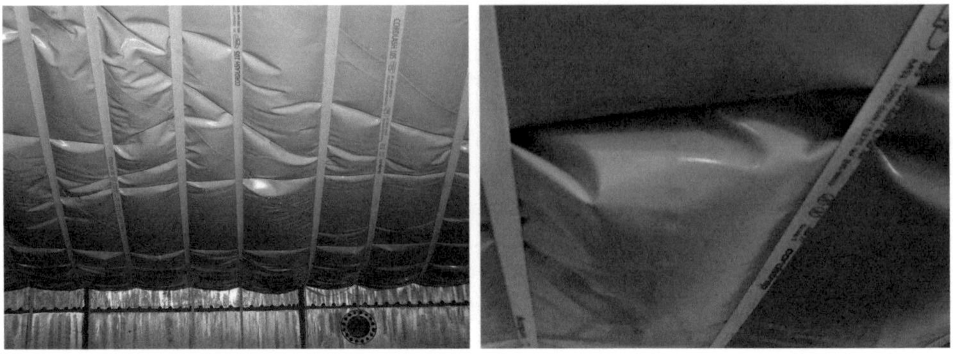

Figure 18.
Depositing the gas membrane onto a belt system [© Rosemarie Wagner].

increase in internal pressure during filling without overpressure protection and at a constant temperature. According to the simplified thermal equation of state, the product of volume and internal pressure is constant for closed volumes [28]. For an internal pressure in the empty tank that corresponds to the ambient pressure of 1013.5 hPa = 101.35 kN/m^2, a volume decrease of 10% during filling leads to an internal pressure of 11.1 kN/m^2, and the gas membrane can be stressed up to failure, depending on the diameter and dome height. To avoid this, pressure relief valves must be installed in the gas chamber and adjusted to the operation-related filling quantity and the volume change.

4.3 Gas production

After sufficient substrate has been filled in and brought to the production temperature required for the bacteria, the fermentation process and gas production begin. Depending on the silage, overproduction occurs in one to four hours, which is up to 50% higher than the daily production [29]. The gas membrane can already be lifted from the support structure by the internal pressure that builds up if it is higher than the internal pressure in the air support space. During gas production, the internal pressure continues to increase, and the gas membrane is in an indifferent state of equilibrium. This results from the internal pressure in the gas space, which is directed in the opposite direction to the dead weight of the membrane with sulphur deposits and internal pressure in the air support space and is equal in amount. External influences such as gusts of wind are transferred to the gas chamber *via* the internal pressure in the air support space, change the equilibrium in which the gas membrane is located and lead to movements. In a radial cutting pattern, the polar cap is heavier than the membrane itself. The reasons for this are the close seam spacing of two adjacent strips and the design of the polar cap with two superimposed, flat and circular membrane surfaces. The more than twice the weight of the polar cap means that it floats much deeper in the gas chamber, **Figure 19.**

In order to examine the behaviour of the gas membrane during the increase of the internal pressure, laser measurements were carried out over the central axes of the cover in the north–south and east–west directions. The curves, which resulted from 24 measuring points for each axis, show that the polar cap has deposited outside the centre point and the influence of the weight of the polar cap on the geometry of the gas membrane. This begins to strain out from the edge of the tank, and the polar

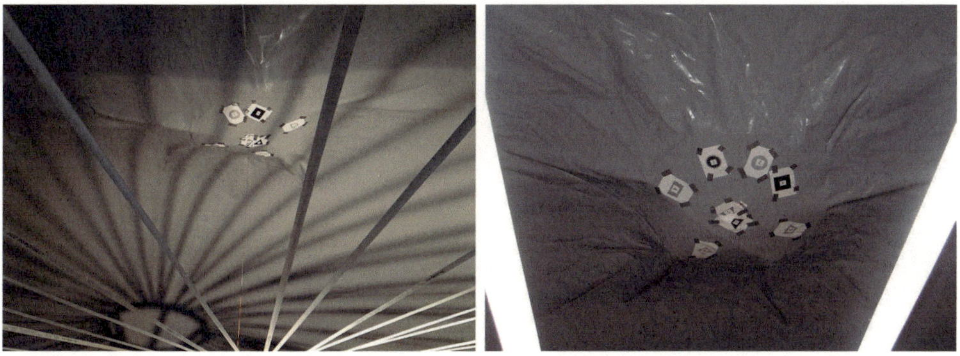

Figure 19.
Geometry of the gas membrane during gas production caused by the weight of the pol cap [© Rosemarie Wagner].

cap hangs much lower. A torus-like geometry is created until almost the final state is reached and the internal pressure in the gas chamber is higher than the weight of the polar cap, **Figure 20**.

The fermentation process is the decisive variable of the biogas yield in m^3 of biogas produced in relation to one kg of substrate and the biogas production rate in m^3 of biogas in relation to the filling volume of the container and day [30].

In order to examine the influence of gas production on the internal pressure in the gas and air support space, gas production was simulated by increasing the internal pressure with a powerful fan. For a cycle of filling, the result is a gas production that corresponds to a volume flow of approx. 550 m^3/h. The inertia of the overall system is shown by the interaction of the internal pressure in the gas and air support space with the increase in the volume flow into the gas space, **Figure 21**. The gas production is similar to the volume flow in the gas space. Even if the volume flow in the air-supported space is constant, the internal pressure rises up according to the volume flow into the gas space.

This also applies after the gas space is filled. The control of the air support pressure to 250 Pa and the controlled internal pressure in the gas chamber with a frequency of the fan of 40 Hz lead to the same behaviour. The decrease in gas pressure at the test tank is explained by leakage on the overpressure valve, and the untightens of fixing to the tank boundary. Gas production is balanced against the atmosphere *via* the air

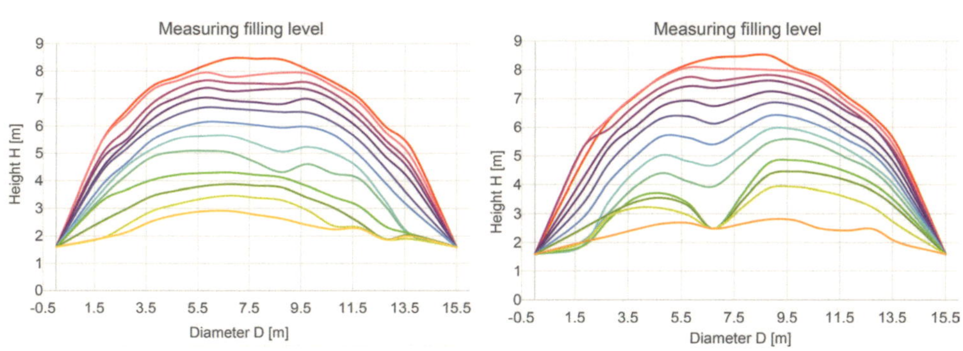

Figure 20.
Measured geometry during filling of the gas space, parallel to the cable winch (left) and perpendicular to it (right) [© Rosemarie Wagner].

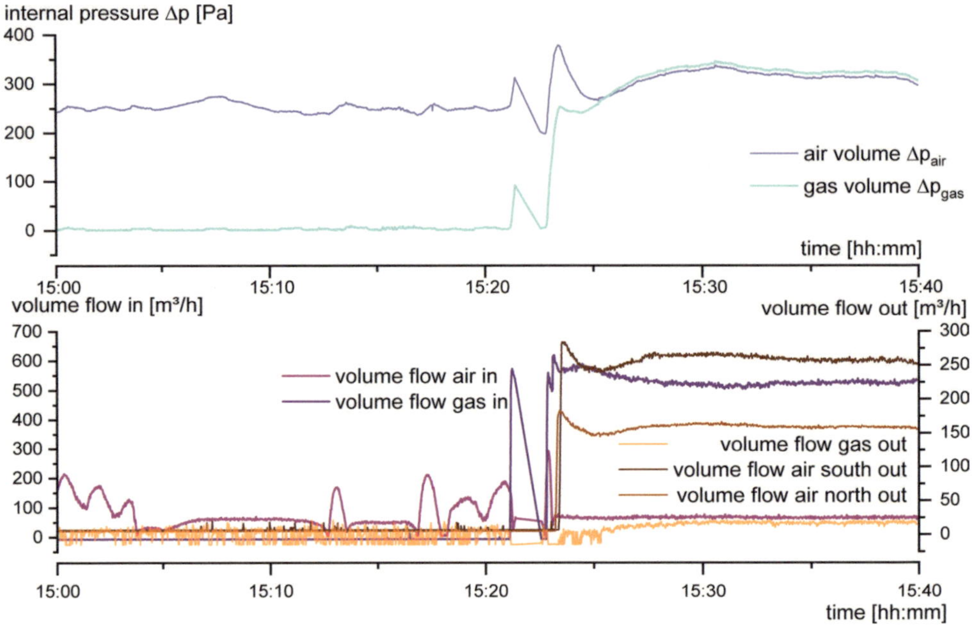

Figure 21.
Internal pressure, volume flow in and out of the two spaces according to the filling of the gas space [© Rosemarie Wagner].

support space with high-volume flows. This shows a direct relationship between air support and gas pressure during gas production.

4.4 Gas extraction

Gas extraction has the consequence that the volume in the air support space drops down as soon as the volume is taken out in the gas space. Volume increase leads to a decrease in internal pressure. In order to maintain the internal pressure in the air support space, almost the volume flow that is taken from the gas space is necessary. If tracking is dispensed with, the same behaviour of the internal pressure can be seen in both chambers. The increase in the volume in the support air space leads to a simultaneous decrease in the internal pressure in the support air space despite the control of the internal pressure by the constant volume flow, **Figure 22**. In order to ensure that internal pressure is maintained for the outer membrane that provides sufficient stiffness against external loads, a sufficient volume flow rate is required into the air support space. This means adjusting the air support volume to the gas volume.

If gas is removed and the membrane is already on the support structure, a negative pressure is created due to the high tightness of the gas chamber when further gas is extracted. With the superposition of the air support pressure and the dead weight, the gas membrane is pressed much more strongly on the support structure. If a belt system is used, the belts are subjected to additional stress. It remains to be seen how large this internal pressure is to be applied for the design of the belts and the centre mast under real conditions. If these situations occur several times, this leads to overstrain of the belts, and the folded membrane is pressed between the belts. The result is damage to the folded membrane at folds and kinks, pushing through membrane between the belts, **Figure 23**.

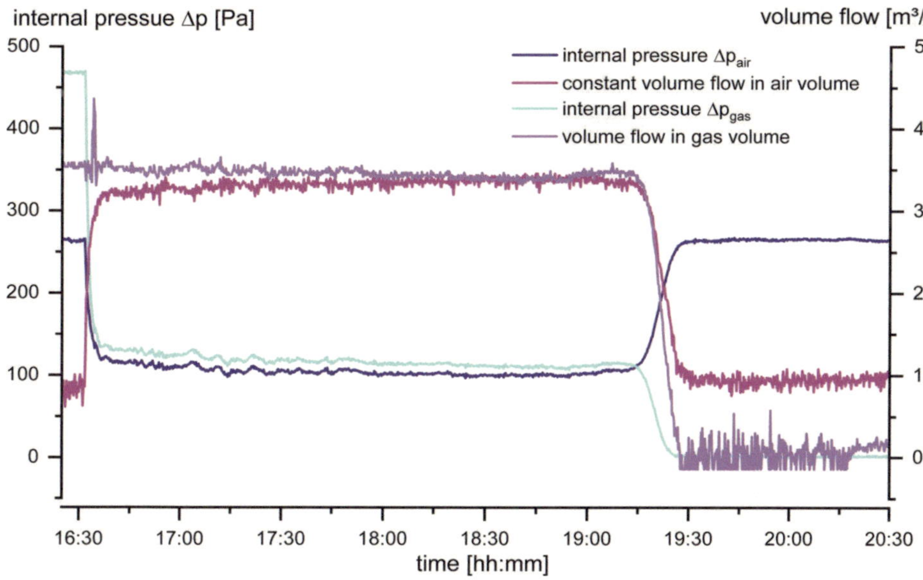

Figure 22.
Gas membrane with higher internal pressure in the support air space than in the gas chamber [© Rosemarie Wagner].

Figure 23.
Gas membrane, loaded with pressure in the air-supported volume and negative pressure in the gas volume, from below (left) [© Rosemarie Wagner] and from above (right) [© Kai Heinlein] the gas membrane.

In addition to inspecting the gas chamber during operation, this is one of the reasons why a plastic net is also placed on the belts today. There is a vacuum protection system for the gas chamber, which is intended to prevent too much stress on the folded gas membrane. The design of the volume flow to compensate for the negative pressure depends on the diameter of the pipe cross-section, the vacuum protection and the pressure in the air-supported volume. For the durability of the gas membrane, it is necessary to ensure a low vacuum. If the system is operated over a day to meet the different demands for electricity, the gas is filled and removed at least once or twice. If there is no adjustment of the internal pressure in the air support space or if it is too sluggish, the outer membrane will flutter when the wind is applied at the same time.

4.5 Full gas volume

Full gas space means either the gas membrane rests against the outer membrane if it is larger or has large expansions as a PE film. If a coated fabric is used, the gas membrane has a larger radius compared to the outer membrane and a small volume of air remains. The support ventilators and the overpressure protections, often exhaust air flaps, have to compensate for the change in internal pressure in the air support space. This is due to fluctuations in ambient pressure, ambient temperature, solar radiation, wind flow and snow. For the full gas space, the internal pressure is higher than in the air support space, and thus, the gas volume is stiffer for wind and snow. Consequently, the gas membrane participates in the load transfer.

5. Measuring the level of filling

The view into the gas chamber is only possible through a small window, which is mounted above the fermentation suspension in the tank wall. Through this opening, it is impossible to observe the gas membrane during filling and emptying over the entire tank. In addition, the internal pressure in the gas space is low for a long time, and only when the gas space is filled above 50% does it rise, and the membrane begins to tighten. For this reason, the filling level of the gas chamber is measured on the systems [31]. Commercially available are cable winches and gas H-metres. For the cable winch measurement, a belt is placed on the gas membrane as centrally as possible in the air support space and secured in position. This is guided through the outer membrane, guided in a weather-protected tube above the edge of the tank and provided with a weight at the free end. The fill level can be read off by changing the length of the belt. The gas H-metre (https://www.seitz-electric.de) determines the level by means of the hydrostatic pressure of a liquid. A hose is filled with liquid and placed on the outside of the gas membrane. The hose extends from the edge of the tank to the polar cap. Depending on the height difference between the two hose ends, a pressure difference occurs. This is converted into an electrical signal and displayed. Both systems work with weight on the gas membrane. Lifting them requires a certain

Figure 24.
Folds in the gas membranes caused by measuring devices, below the gas membrane (left), from above (right) [© Kai Heinlein].

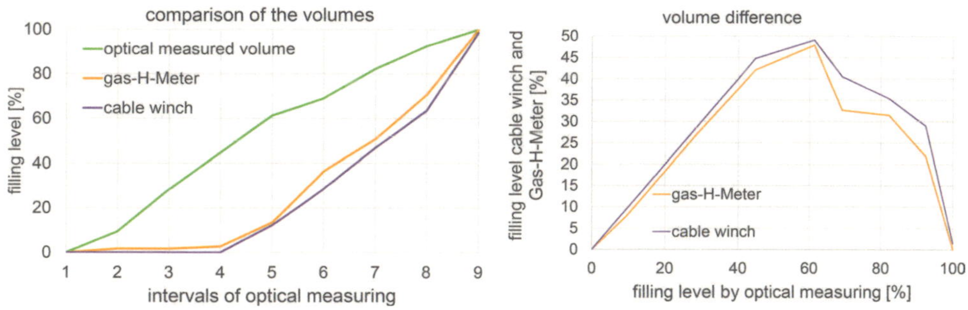

Figure 25.
Filling level related to different measures of the gas [© Rosemarie Wagner].

internal pressure. As long as this is smaller, there is a fold of the membrane at the point where the devices are positioned on the gas membrane, **Figure 24**.

The tension of the cable winch caused by its weight pulls the membrane downwards till the gas pressure is high enough to rise the weight. It happens the same as the membrane is pressed on the belt system. Along the cable, winch wrinkles occur regularly. The folds and kinks are also reasons to damage the coating and lead to a higher gas emissions. The wrinkles are related to the height of the membrane. The higher it is, the deeper and more crumbled folds and kinks can be found.

The same wrinkle occurs with the liquid-filled hose of the gas H-metre. This means that the level can only be measured when the gas membrane is tensioned, and the internal pressure is reached, which is used for the design of the membrane. The fill level is displayed up to 50% lower than it actually is due to the folds and geometry of the gas membrane, **Figure 25**. The optical measuring is derived from the curves given in **Figure 20**. It is assumed that the geometry between the two lines of measured points is the same.

6. Conclusion

The content deals with the applications, advantages and disadvantages of membrane covers for the storage of biogas. Due to the geometry, there are differences between mechanically pretensioned and internal pressure-supported covers in the storage volume, use and stresses. Mechanically tensioned covers do not require a substructure; the central mast to support the membrane requires a foundation because the entire load on the membrane is transferred to it via compression. The covers should have a two-axis curvature to be more stable against wind-induced fluttering and to avoid water sags. Covers supported by internal pressure require a substructure because, without internal pressure, they are folded on top of it. Layered covers are more complex to manufacture than single-layer covers. Separating the storage of gas and removing external influences such as wind and snow increases durability and reduces leakage compared to single-layer covers. For the membrane materials, the requirements for the special conditions, such as the precipitation of sulphur, the basic and acidic environment, the surface quality and the buckling behaviour, must be considered. It must be considered that the strain and stress behaviour lead to permanent elongation after the first load cycles. This results in different geometries and different stresses in the membranes. The deformations also differ from the tensions in the warp and weft directions. In two-layer internal pressure-supported covers, the gas volume, the gas pressure, the air support volume, and the internal pressure interact in the air support volume. This must be considered, particularly in operating conditions. The geometry of the gas membrane is in balance with its own weight, including condensation and sulphur deposits, as well as the gas pressure and air pressure in the air support space. Higher self-weight at the pole cap or measurement device for the fill level results in different geometries and other volumes. These differ from the volume of a spherical segment.

Author details

Rosemarie Wagner* and Kai Heinlein
KIT Karlsruhe Institute of Technology, Karlsruhe, Germany

*Address all correspondence to: rosemarie.wagner@kit.edu

References

[1] Wagner R. Bauen mit Seilen und Membranen. Berlin: Beuth Verlag; 2016

[2] Deutsche Vereinigung für Wasserwirtschaft, Hrsg., DWA-Merkblatt 377 (DVGW-Merkblatt G436-1) Biogasmembranspeichersyssteme über Behältern, Hennef; 2022

[3] Ströbel D, Holl J. On the static calculation of biogas containers with radial and parallel cutting pattern. In: Lázaro C, Bletzinger K, Oñate E, editors. Form and Force. Barcelona: International Centre for Numerical Methods in Engineering (CIMNE); 2019. pp. 380-387

[4] Eder B, Schulz H, Krieg A, Mitterleitner H. Biogas-Praxis: Grundlagen, Planung, Anlagenbau, Wirtschaftlichkeit, Beispiele. Rastede: Ökobuch; 2007

[5] Roedel W, Wagner T. Physik unserer Umwelt: Die Atmopsäre. Berlin: Springer-Verlag GmbH Deutschland; 2017

[6] Johann J. "Wie gasdurchlässig sind Foliendächer? Biogasspeichermembranen: Folientypen im Test," 30 08 2010. [Online]. Available from: https://www.ikz.de [Zugriff am 11 10 2024]

[7] DIN EN ISO 52-17:2012. Kunststoffe - Bestimmung der Zugeigenschaften (ISO 527-1:2012). In: Technisches Komitee ISO/TC 61 "Plastics"; Technisches Komitee CEN/TC 249 "Kunststoffe", eds. Berlin: Beuth; 2012

[8] Moneke M. Kunststoffwerkstoffe. München: Carl Hanser Verlag; 2022

[9] DIN EN 17117-1: 2018. Mit Kautschuk oder Kunststoff beschichtete Textilien - Mehanische Prüfverfahren unter biaxialen Spannungszuständen (EN 17117-1:2018). Technisches Komitee CEN/TC 248 "Textilien und textile Erzeugnisse", eds. Berlin: Beuth; 2018

[10] Göschel U. Thermo-Mechanische Eigenschaften. In: Elsner P, Eyerer P, Hirth T, editors. Kunststoffe Eigenschaften und Anwendungen. Heidelberg: Springer Verlag; 2012. pp. 130-146

[11] Pauly S, Becker A. Bestimmung Methandurchlaääsigkeit. Wiesbaden: Fachlaboratorium für Permationsprüfung; 2009

[12] Röthemeyer F, Sommer F. Kautschuk Technolgie Werkstoffe - Verarbeitung - Prdoukte. Münnchen: Carl Hanser Verlag; 2013

[13] Johann J, Gramatte W. Biogasspeichermembranen (Elastomer). Germany: Groß-Umstadt; 2009

[14] Büeler E. CH4-Emissionen bei EPDM-Gasspeichern und deren wirtschaftlichen und ökologischen Folgen. Swiss: Bern; 2011

[15] Elsner P, Eyerer P, Hierth H. Kunststoffe Eigenschaften und Anwendungen. Heidelberg: Springer Verlag; 2012

[16] Affolter S. Langzeitverhalten von Thermoplasten. Swiss: Buchs; 2004

[17] Baur E, Drummer D, Osswald TA, Rudolph N. Saechteling Kunststoff-Handbuch. München: Carl Hauser Verlag; 2022

[18] K. f. Anlagensicherheit. TRAS 120: Sicherheitstechnische Anforderungen an Biogasanlagen, 2022

[19] Harvie G. An Investigation into the Thermal Beaviour of Spaces Enclosed

by Fabric Membranes. Welsh School of Architecture: University of Wales College of Cardiff; 1996

[20] Wagner R, Heinlein K. Innendruck in zweilagigen, textilen Abdeckungen über Biogasspeichern. Der Bauingenieur. 2022;**97**(12):393-402

[21] Wechselberger V, Reinet T, Yngvesson J, Scharfy D, Schuetz C, Huber-Humer M, et al. Methane losses from different biogas plant technologies. Waste Management. 2023;**157**:110-120

[22] Bakkaloglu S, Lowry D, Fisher RE, France JL, Brunner D, Chen H, et al. Quantification of methane emissions from UK biogas plants. Waste Management. 2021;**124**:83-93

[23] Müller-Rochholz J, Bronstein Z. Einfluß von Zugebanspruchungen auf das Hydrolysverhalten von Polyester. Stuttgart: Fraunhofer IRB-Verlag; 1994

[24] Schröder HF. Ermittlun des Einflusses der alkalischen Hydrolyse für die Langzeitbeständigkeit von hochfesten Polyestergarnen für Geotexilien. Berlin: Bundesanstalt für Materialforschung und-prüfung; 1999

[25] DIN EN ISO 1421:2018. Mit Kautschuk oder Kunststoff beschichtete Textien - Bestimmung der Zugfestigkeit und der Bruchdehnung (EN ISO 1421:2018). Technisches Komitee ISO/TC 45 "Rubber and rubber products", Technisches Komitee CEN/TC 248 "Textilien und textile Erzeugnisse", eds. Berlin: Beuth; 2018

[26] Wagner R, Heinlein K, Sum B. Erforschung neuer Lösungen für Biogasspeichersysteme. Karlsruhe: Eigenverlag; 2019

[27] Weiland P, Rieger C, Ehrmann T. Biogas-Anlagen 12 Datenblätter. Gülzow: Fachagentur Nachwachsende Rohstoffe e. V. (FNR); 2004. pp. 7-29

[28] Frohn A. Einführung in die technische Thermodynamik. Stuttgart: AULA-Vrlag Wiesbaden; 1989

[29] Paterson M, Hülsemann B, Roth UDWV, Wirth B, Bidlingmaier C, Hartmann S, et al. BIOGAS PROGRESSIV—Zukunftsweisende Strategien für landwirtschaftliche Biogasanlagen. Kuratorium für Technik und Bauwesen in der Landwirtschaft e.V. (KTBL). 2023

[30] Wesselak V, Schabbach T, Link T, Fischer J. Handbuch Regenerative Eneergietechnik, 3 Hrsg. Berlin: Springer Vieweg; 2017

[31] Effenberger M, Kissel R, Aschmann V, Herb C, Helm MMJS, Maciejeczyk M, et al. Empfehlungen für die messtechnische Ausstattung landwirtschaftlicher Biogasanlagen. Freising-Weihenstephan, Germany: ALB Bayern e.V; 2021

Chapter 5

Biomass and Biogas to Hydrogen

Wilson Hago and John O'Connell

Abstract

Biomass holds immense promise for augmenting existing energy reserves by converting it to low-carbon hydrogen for direct use in transportation, semiconductor manufacture, power plant cooling, and as a reactant for ammonia, foods, and high-value chemicals. In particular, biogas extracted from diverse sources such as agricultural, animal, and municipal waste, and wastewater treatment can serve as feedstock for hydrogen production. This chapter discusses contemporary technologies for hydrogen production from biomass by comparing various feed sources and production methods, as well as exploring the economic, greenhouse gas emission, and societal implications of using biomass as feedstock for hydrogen production. A process under development by the authors is also described.

Keywords: biomass, biogas, hydrogen, biomethane, methane, biological method, thermochemical method

1. Introduction

Because it can be utilized in many ways without significant greenhouse gas (GHG) emissions, hydrogen is considered essential to the transition from materials and energy using carbon-based sources-such as coal and oil-to carbon-neutral production. Hydrogen is not found naturally, so raw materials containing it, such as water and methane, along with significant energy, are needed to produce it. Currently, nearly all the 100 Mt. per year produced is used in: (1) pure form for ammonia production and oil refining, and (2) mixed with other gases for chemical and steel manufacturing. The present amounts used in heavy industry, transportation, and energy storage, where its contribution to the transition would be most effective, are less than 1% of the total demand. Further, only about 1 Mt. per year comes from the low-GHG-emitting processes of water electrolysis, biomass, carbon capture utilization, and storage (CCUS). The overwhelming majority is from fossil sources, principally natural gas. If the world is to achieve Net Zero Emissions (NZE) by 2050, low-emissions hydrogen (green hydrogen) production will need to be more than 60 Mt. per year [1].

Many processes have been developed for generating hydrogen. Refs. [2–17] provide details of the various methods, including those with fossil fuel feed, principally natural gas, and those with renewable feed, such as biomass and water. The technologies include natural gas decomposition, biomass conversion, electrolysis, thermolysis, and biological reactions including bioelectrolysis.

Implementation of large-scale facilities for hydrogen production will not only depend on the technologies available but also the economics of obtaining raw materials, construction and operation of facilities, storage, and delivery. To indicate the present variations, **Table 1** gives some estimates of the costs for hydrogen with various technologies. AlHumaidan et al. [14] also describe in detail the current status of hydrogen storage and delivery, as well as give a thorough discussion of the current status of carbon capture methods, utilization, and storage. Singh et al. [9] describes the Technology Readiness Levels (TRL) of the available technologies, along with Life Cycle Analyses (LCA) of the methods.

The focus of the present chapter is on processes with renewable biomass feeds. Rey et al. [10] give a very recent and comprehensive analysis of the technology and economic aspects of such green hydrogen production.

Among the processes for green hydrogen production, using biomass as the source material has many attractive features [2–4, 7, 9–11, 16, 17, 19]. Its use is carbon-neutral because the GHG emissions are essentially the same as those from waste decomposition or even less if some of the products that would emit CO_2 are sequestered. Further, environmental problems such as air and water pollution from animal and agricultural waste are somewhat alleviated by their collection for conversion. Finally, the energy requirements for biomass conversion can be less than electrolysis or CCUS [4, 7].

While the economics of biomass conversion are generally favorable, there are serious technical challenges for large-scale implementation of biomass technology: biomass availability, gasifier efficiency, integration into existing energy systems, and disposal of the byproducts, principally solid carbon [2, 4].

This chapter is organized to first introduce different technologies associated with biomass conversion to hydrogen and then present their technoeconomic results with a particular example process.

2. Hydrogen production from biomass

Figure 1 is a block diagram of the common methods to obtain H_2 from biomass. The principal categories are biological and thermochemical, where the former involves treatment with aqueous microorganisms to degrade biocompounds and produce hydrogen along with other compounds. The latter uses chemical reactions to convert the gaseous products of biomass degradation, normally CH_4, into hydrogen and byproducts such as carbon and carbon oxides.

The source biomass materials are usually lignocellulosic residues from agriculture or forestry, animal wastes, sewage, and municipal and industrial waste [6, 7, 10, 11]. Pretreatment of the biomass is necessary to eliminate undesirable chemical and microbial components, to resize solids, and to remove water by drying [19]. In all

Technology	Feed	Cost, \$USD/kg H_2	Reference
Natural Gas Decomposition	Natural Gas	\$0.77–\$2.10	[6, 17, 18]
Biomass conversion	Biomass	\$1.30–\$6.60	[6, 17]
Electrolysis/Thermolysis	Water	\$2.35–\$7.70	[6, 7]
Biological	Aqueous chemical, biomass	\$1.56–\$11.39	[6, 7, 17]

Table 1.
Estimates of production costs for different hydrogen production technologies.

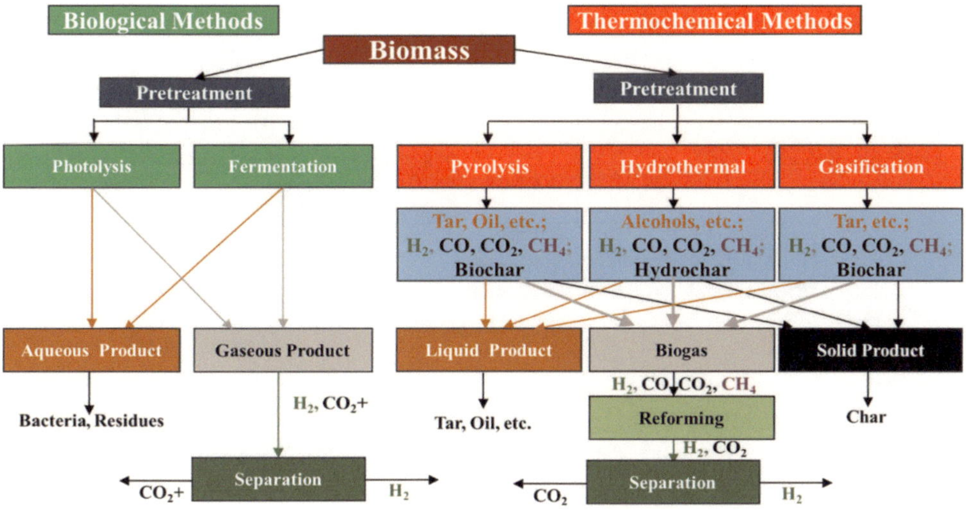

Figure 1.
Methods for biomass conversion to hydrogen. After Ref. [7].

cases, the gaseous product is a mixture of hydrogen-containing substances, such as hydrogen or methane, along with other components, such as carbon oxides.

Biological processes yield H_2, CO_2, and other substances. Purification is principally the separation of dilute hydrogen. The chemistries, bacteria utilized, operating conditions, and hydrogen yield are tabulated in Refs. [12, 13]. There are also anaerobic light, anaerobic dark, and integrated process systems (indicated in **Figure 1** as fermentation). Thermochemical processing (biogas) yields H_2, CO, CO_2, and CH_4. This is followed by reforming reactions to yield only H_2 and CO_2 for purification.

Electrochemical processing is normally associated with water splitting to make hydrogen and oxygen. However, there have been recent investigations into using electrodes with biomass [6, 11, 13].

Some details of these processes are discussed in Section 2 with a focus on the processes for treating the gaseous products.

2.1 Thermochemical methods

Commonly used thermochemical methods include pyrolysis, hydrothermal treatment, and gasification.

- Pyrolysis involves heating pretreated biomass to high temperatures (300–800°C) in the absence of O_2. The products are gaseous H_2, CH_4, CO_2, CO, and H_2O. Liquid tars and oil are found as well as solid char. Higher temperatures run faster and yield less char. **Figure 2** illustrates the process configuration for traditional pyrolysis where the reforming step of **Figure 1** is included. Heat integration is possible, but not shown. A novel approach to pyrolysis is plasma pretreatment followed by reforming [20, 21]. Because lignocellulose is quite stable, it may not be fully degraded by traditional pyrolysis. Plasma is more able to break C-C and C-H bonds. Further, the treatment not only reduces the lignin product, it increases the sugar availability [15].

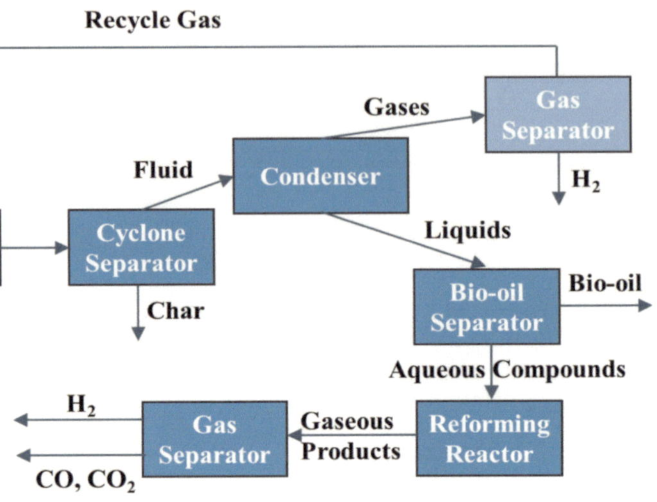

Figure 2.
Pyrolysis process diagram. After Ref. [10].

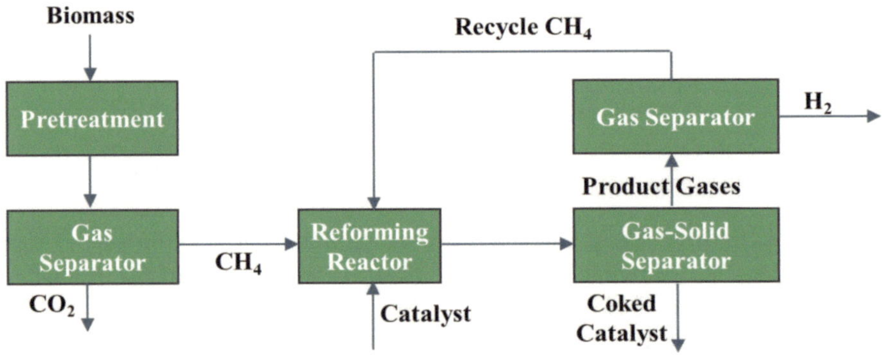

Figure 3.
Flow diagram of a general gasification process.

- Gasification involves very high temperatures (800–1100°C) to crack the biopolymers to gaseous products in the presence of a catalyst. **Figure 3** shows the flow diagram which includes the reforming step of **Figure 1**. Pretreatment must remove all components except CO_2 and CH_4 and condition the feed for gas separation of the CO_2 byproduct and the CH_4 reactant. The reactor can be a moving bed or fluidized [22] so that fresh catalyst can be fed along with the CH_4, and the coked catalyst can be removed and separated from the product gas mixture. Since the extent of the CH_4 decomposition reaction is not complete, unreacted CH_4 must be recycled after gaseous separation. If the coked catalyst is sequestered such as by amending soil, the GHG emissions can become negative.

A technoeconomic example process for biogas conversion to H_2 is given in Section 3 with its Life Cycle Analysis (LCA) in Section 4.

- Hydrothermal processing

Hydrothermal processes involve high-temperature, high-pressure water as both reactant and reaction medium. There are three different approaches involving carbonization (HTC), liquefaction (HTL), and gasification (HTG) [7] characterized by their temperature ranges. HTG uses supercritical water (T > 374°C, P > 22.1 MPa) while the others operate at lower temperatures and pressures. While the process conditions are extreme and the process configuration complex, there are a number of advantages to this treatment. In addition to competitive costing, a wider variety of biomass can be handled, it need not be dried, and the products can be of high-value with less tar being produced [7]. The output of hydrothermal processing requires reforming.

2.2 Reforming

The goal of reforming is to generate more H_2 from the biogas components either by reacting with H_2O (wet reforming), CO_2 reforming (dry reforming), or by decomposing the CH_4 (catalytic decomposition). It is a standalone process when fossil fuels are the feed. The reaction mechanisms are detailed in Ref. [23, 24].

2.2.1 Wet reforming

Reaction of methane with steam (SMR) is the most mature and widely used process for hydrogen manufacture [3, 13, 14, 24]. It is applicable for both fossil fuel and biogas feeds which produce methane. The temperature is high (700–1000°C) and the process involves a catalyst, usually nickel-based.

$$\mathrm{CH_4} + H_2O \rightarrow \mathrm{CO} + 3\mathrm{H_2}\ \text{(Endothermic)} \tag{1}$$

$$\mathrm{CO} + H_2O \rightarrow \mathrm{CO_2} + H_2\ \text{(Exothermic)} \tag{2}$$

Reaction (2) is the water-gas shift reaction (WGS). A variation on SMR is partial oxidation (POX) where the feedstock is partially oxidized *via* reactions (3) with reaction (4) in place of reaction (2) at a very high temperature (1150–1500°C).

$$\mathrm{CH_4} + \tfrac{1}{2}\, O_2 \rightarrow \mathrm{CO} + 2\mathrm{H_2}\ \text{(Endothermic)} \tag{3}$$

$$\mathrm{CH_4} + O_2 \rightarrow \mathrm{CO_2} + 2\mathrm{H_2}\ \text{(Endothermic)} \tag{4}$$

Another variation is autothermal reforming (ATR) which is a combination of SMR and POX with both O_2 and H_2O added to the feed and run at intermediate temperature (900 + 1100°C) and at moderate pressure (80–100 bar) [3, 9, 14].

By varying the relative amounts of CH_4, H_2O, and O_2, in addition to temperature and pressure, the product compositions will change. While the objective is normally to produce H_2, some systems are set to yield higher amounts of CO to produce syngas that can be used for the synthesis of high-value chemicals [9, 14].

Note that all these processes require added heat, traditionally from the combustion of fuels. It is stated that to produce 1000 kg of H_2 by SMR requires 5.2 MWh of heat energy, 2000 kg CH_4, and 4500 kg H_2O, in addition to contributions from an air separation unit (ASU) for the O_2. The GHG produced is 5500 kg CO_2. Using concentrated solar energy for the heat source is being investigated with a significant reduction in the fuel energy requirement being expected [3, 12].

2.2.2 Dry reforming (DMR)

In addition to processes for reforming methane with water, there is dry reforming of methane with CO_2 *via* reaction (4) [14, 18]. It is run at 700–900°C with a CO_2/CH_4 ratio of 1 to 1.5. Also occurring will be some extent of the reverse of reaction (2), decreasing the production of H_2, but increasing the amount of CO for syngas. Both of these reactions are endothermic, so significant heat is required. It is reported that the Ni-based catalysts can become coked from CH_4 decomposition.

2.2.3 Methane decomposition

The reforming reaction (sometimes referred to as methane pyrolysis), is

$$CH_4(g) \rightarrow C(s) + 2H_2 \tag{5}$$

Mechanisms and catalysts for this reaction are reviewed in Ref. [23–25]. Many supported metal catalysts and carbon forms are mentioned, though biochar was not, even even though it has been successfully used [22, 24, 26]. The carbon product deposits on the catalyst, including inside the pores, often limiting the reaction rate and extent.

2.3 Biological conversion technologies

Comprehensive reviews of hydrogen generation using biological methods are given in [7, 11, 12]. **Figure 4** shows the methods organized as in **Figure 1**.

Biological hydrogen production involves microorganisms that metabolically produce hydrogen gas by fermentation or photosynthesis. The methods are anaerobic and

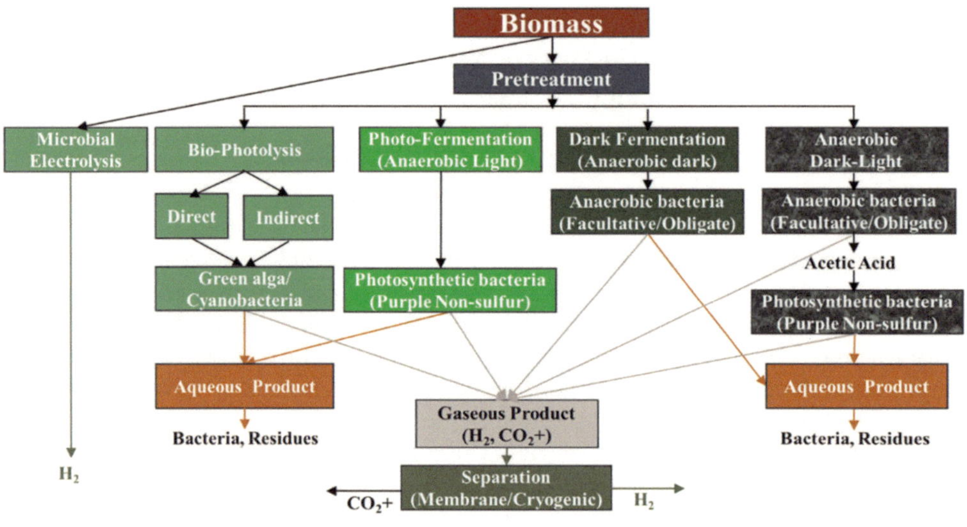

Figure 4.
Biological methods to produce hydrogen from biomass. After Ref. [7].

may involve light or darkness. Photo-fermentation and bio-photolysis (direct or indirect) are light-dependent methods using algae or microalgae, while dark fermentation uses bacteria. Hydrogen production from algae, microalgae, and bacteria produces high yields at temperatures between 20 and 45°C at a lower cost than other production methods. However, some secondary reactions can occur, such as oxygen from photosynthesis and organic acids from organic compound decomposition. These limit hydrogen production to lower yields. Anaerobic dark–light integrated systems can mitigate these issues while increasing efficiency, production rate, and hydrogen yield [7].

2.3.1 Bio-photolysis

Hydrogen production *via* bio-photolysis uses photosynthetic oxygenic microorganisms, including green microalgae (*e.g.*, *Chlamydomonas reinhardtii*, *Chlorella vulgaris*, *Scenedesmus obliquus*, and *Chlorococcum minutum*) and cyanobacteria (*e.g.*, *Nostoc* sp., *Cyanothece* sp.). Solar energy causes photosynthesis and transfers electrons for hydrogen with hydrogenase as a catalyst. Direct bio-photolysis (DbP) is when photoautotrophic organisms convert water into H_2 and O_2 by the simultaneous reactions

$$H_2O \overset{\text{Light}}{\rightarrow} 2\text{H}^+ + 2\text{e}^- + ½\, O_2 \quad (6)$$

$$2\text{H}^+ + 2\text{e} \overset{-\text{Hydrogenase}}{\rightarrow} 2\text{H}_2 \quad (7)$$

During indirect bio-photolysis (i-DbP) cyanobacteria synthesize carbohydrates *via* reaction (8) and cell metabolism by microalgae forms hydrogen via reaction (9)

$$6\text{CO}_2 + 6\text{H}_2O \overset{\text{Light}}{\rightarrow} C_6H_{12}O_6 + 6O_2 \quad (8)$$

$$C_6H_{12}O_6 + 6\text{H}_2O \overset{\text{Light}}{\rightarrow} 12H_2 + 6O_2 \quad (9)$$

Rey et al. [7] list the microorganisms, substrates, operating conditions, and H_2 yield for some systems. The substrates include glucose, glycerol, corn stalk, and starch. Glucose is the common substrate for bio-photolysis. Also given are the variables that affect H_2 yield including temperature, pH, O_2 content, light intensity, nitrogen and sulfur concentrations, added organic carbon, cell density, and culture age.

2.3.2 Fermentation

There are several different routes for producing H_2 *via* fermentation. Photo-fermentation generates H_2 by photosynthetic bacteria, primarily purple non-sulfur (PNS) bacteria, with several steps catalyzed by nitrogenase and light. Yields are affected by anaerobic conditions, light intensity, temperature, pH, wavelength, and substrate concentration type. Oxygen inactivates the nitrogenase so anaerobic conditions are necessary. The best substrates are short-chain fatty acids and volatile fatty acids (e.g., acetate, butyrate, lactate, malate, etc.). Sugar substrates lower the hydrogen production.

Dark fermentation can be run with a wide variety of anaerobic microorganisms in the absence of light and, depending on the microorganism, at temperatures of 25–45°C,

45–80°C, or T > 80°C. Typically, the substrate, such as carbohydrates or other feedstocks, has high glucose. There are several metabolic pathways that have different H_2 yields. While the theoretical yield of H_2 from glucose *via* acetic acid is 4 mol H_2/mol glucose, the actual yield can be as low as 50% of the theoretical. Again Rey et al. [7] list many different inocula, substrates, operating conditions, maximum H_2 yields, and references as well as the variables that affect H_2 yield.

There is also the possibility of combining photo-fermentation with dark fermentation in either a one-step or the preferred two-step process [12]. First, dark fermentation bacteria generate some H_2 and intermediate products such as acetic acid. Then, photo-fermentative bacteria convert the intermediates into H_2 and CO_2. The theoretical yield for this process would be 12 mol H_2/mol substrate(glucose) but approaching this requires pretreatments of concentration, composition, and pH.

2.3.3 Microbial electrolysis

Noori et al. [13] describe in detail the possibilities and issues in the production of H_2 with microbial electrolysis cells (MEC). In particular, unknowns about inocula and reactor design have limited the scaling of the technology, but the analysis suggests ways to overcome these difficulties. Unlike water electrolysis where O_2 is produced, the MEC feedstock is oxidized to produce H^+ ions at the anode in the presence of exoelectrogenic bacteria. The efficiency can be high for the mixed cultures of sediments, wastewater, and microbial mats. However, while some microbes produce H_2, others consume it.

A list of advantages and challenges for all the hydrogen generation processes discussed in this section is given in **Table 2**.

Process	Advantages	Challenges
Steam Reforming	No O_2 needed; known technology	High temperatures, catalyst regeneration; CO_2 produced
Gasification	Recycles wastes, known technology; high efficiency; no O_2	High temperatures; catalyst coking; CO_2 produced; high reactor cost
Pyrolysis	Biomass variation, known technology; recycles wastes	Tar, char, & CO_2 produced; catalyst regeneration; high reactor cost
Direct bio-photolysis	High efficiency; recycles wastes; mild conditions	Sensitivity to O_2; low catalyst availability
Indirect bio-photolysis	Separate H_2 and O_2 production, use both hydrogenase and nitrogenase	Low efficiency; environmental sensitivity
Photo-fermentation	Wide range of substrates; mild conditions; recycles wastes	High energy use; sensitivity to O_2
Dark fermentation	Continuous process; wide range of substrates; mild conditions	Low efficiency; H_2 reversal; volatile effluent
Dark–Light fermentation	High biomass conversion; H_2 yield	Capital intensive; specialized infrastructure
Microbial Electrolysis	No separation is needed; low electricity	High-cost catalyst; pH sensitivity; requires simpler feeds

Table 2.
Advantages and challenges of hydrogen generation processes. After Refs. [6, 7].

3. Technoeconomic analysis (TEA) of biogas to hydrogen conversion: The Hago energetics case

The sections above describe the technologies for conversion, considerations for feed and reaction, as well as processing and life cycle analysis for biogas to hydrogen. This section illustrates a particular system that has been studied by process simulation and laboratory experiments and is now being constructed at the pilot scale: The Hago Energetics process.

The path through **Figure 1** that applies to this process is pretreatment to gasification to biogas to reforming to separation. A block diagram of the system is shown in **Figure 5** and an AspenPlus® flowsheet is shown in **Figure 6**. Biowaste, mainly from cow manure, is fed to a digester to generate raw biogas which is upgraded to with biomethane separated from an off-gas product of mainly CO_2. The methane is mixed with recycled methane and biochar. After heat exchange, the methane is fed to a

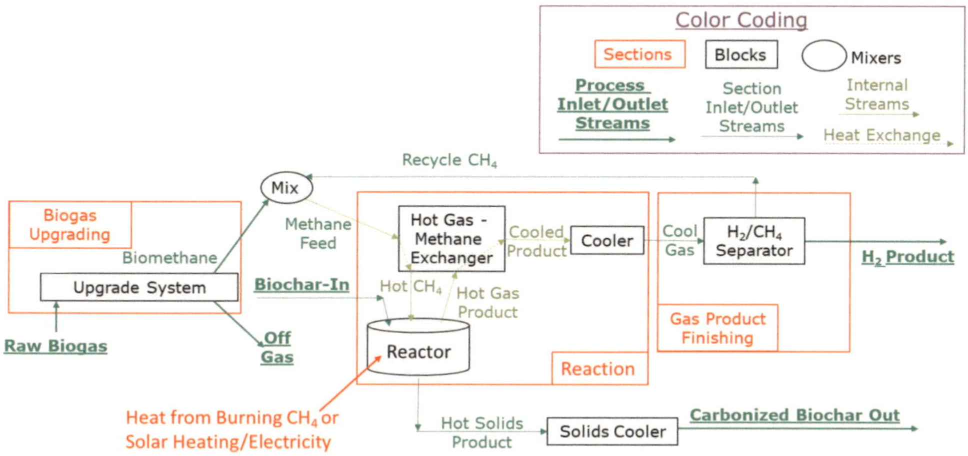

Figure 5.
Block diagram of Hago energetics process for biogas to hydrogen.

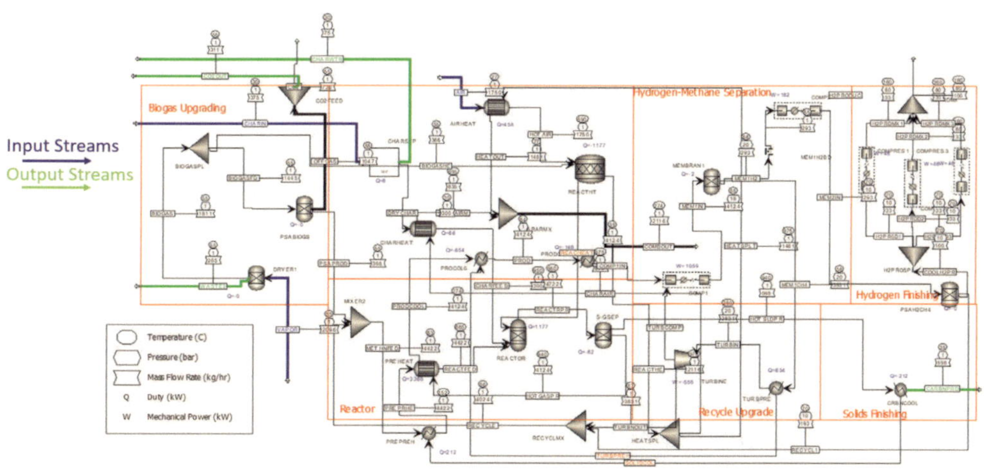

Figure 6.
AspenPlus® flowsheet of Hago energetics process for biogas to hydrogen.

Case	Case 1	Case 2	Case 3	Case 4	Case 5	Case 6	Case 7
Item	**Base**	**Subsidy**	**Subsidy**	**Feed**	**Feed**	**Price**	**Price**
Installed Capital, $MM	$26.4	$26.4	$26.4	$26.4	$26.4	$26.4	$26.4
Fixed Costs, $MM/yr	$4.60	$4.60	$4.60	$4.60	$4.60	$4.60	$4.60
Total Feed Cost, $/kg H_2	$4.89	$4.89	$4.89	$3.91	$5.87	$4.89	$4.89
Biogas Cost, $/kg H_2	$2.85	$2.85	$2.85	$2.28	$3.42	$2.85	$2.85
Utilities, $/kg H_2	$0.55	$0.55	$0.55	$0.55	$0.55	$0.55	$0.55
Gov't Subsidy, $/kg H_2	$13.00	$0.00	$6.50	$13.00	$13.00	$13.00	$13.00
Selling Price, $/kg H_2	$13.40	$13.40	$13.40	$13.40	$13.40	$10.72	$16.08
Sales Revenue, $MM/yr	$9.55	$9.55	$9.55	$9.55	$9.55	$7.64	$11.46
Net Earnings, $MM/yr	$6.68	-$0.45	$3.11	$7.07	$6.29	$5.67	$7.98
Return on Investment, %	25.3%	−1.7%	11.8%	26.8%	23.9%	21.6%	30.1%

Table 3.
Economic analysis of Hago process.

moving bed reactor producing a hot gas product of hydrogen and methane and a hot solids product of char with carbon deposited on it. The reactor can be heated by any of several methods, including solar electricity. For the present case, part of the biomethane is combusted for reactor heating. After heat exchange with the inlet methane, the product gas is further cooled and then the H_2 product is separated from the unreacted CH_4 by membrane filtration and compressed if desired.

An economic analysis of the process has been completed with the capital and operating costs listed in **Table 3**. The values were obtained using an .xlsx template from Downey and Vrana based on the economics in the text by Seider et al. [27]. The H_2 product conditions are 100 kg/h at 900 bar, so a compressor and cooler are included in the equipment. Several cases are explored. The base is Case 1. Government subsidies of $0.00 (Case 2), $6.50 (Case 3), and $13.00 (Case 1) have been used. Costs of biogas (Cases 4 and 5) and the selling price of hydrogen (Cases 6 and 7) have been varied ±20% to determine the sensitivity of the ROI to feed prices. Changes in selling price have a greater effect on profitability than biogas costs. No credit is earned from the CO_2 produced, but carbonized char is marketed for $0.33/kg H_2. Returns on investment (ROI) are in the range of −1.7 to 30.1% with positive returns only if there is a government subsidy.

4. Life cycle analysis of biogas to hydrogen conversion

Life Cycle Analysis (LCA) is a powerful tool to gauge the effectiveness of an approach toward assessing the overall emissions impact of a technology or approach. It provides a basis for comparison between different technologies and processes that may use a wide range of energy and mass inputs. The starting point of a life cycle analysis is the choice of boundaries for the process under consideration, which limits estimates of the embodied energy of the inputs vs. operational energy needed for the various processes in the system. A mass and energy flow for the basic elements of a process then sets the stage for an estimate of the total emissions out of the system, as well as emissions from chemical or transformational changes within the system. It should be noted that LCA estimates are highly specific to site location and are affected by such local considerations as energy generation, feedstocks used in biomass generation, and transport of inputs and outputs.

Standardized methodologies exist to execute LCA calculations. The international standards are ISO 14040 and 14044. The most comprehensive analysis addresses cradle-to-cradle emissions. Other types of analyses include cradle-to-gate, and gate-to-gate emissions, where gate refers to the production factory boundary. In the case of transport fuels, the reference name is wells-to-wheels analysis. Software for performing LCA analyses includes Microsoft Excel, GaBi [28], SimaPro [29], and GREET [30].

A review of LCA analyses of biogas production in Europe was performed by Hijazi et al. [31]. This review examined 16 LCA studies, covering a wide variety of feedstocks and locations. It was noted that the primary use of the biogas was for electricity production and/or heating.

4.1 Hago Energetics example

The following LCA calculations and results are provided as an example of a biogas-to-hydrogen pyrolysis-based process. The LCA estimate for the Hago Energetics case was performed by third-party firm EcoEngineers [32]. An estimate the Carbon Impact (CI) was developed from hydrogen produced in a proposed biogas to hydrogen project in a dairy farm in the Central Valley of California, USA. Estimates are based on the technoeconomic analysis presented in Section 3. Data was processed with model inputs for the California Air Resources Board Tier 1 CI Calculator for Hydrogen and California GREET 4.0 model. **Table 4** summarizes the project's key parameters based on the above documents and communication with Hago staff.

Three scenarios were analyzed:

- Scenario 1: Gaseous hydrogen (G.H2) production with transport *via* tube trailer truck to the nearest hydrogen refueling station 27 miles away

- Scenario 2: Liquid hydrogen (L.H2) production with transport *via* tanker truck to the nearest hydrogen refueling station 27 miles away

- Scenario 3: Liquid hydrogen (L.H2) production with transport *via* rail 281 miles away (to a nearby port)

The results for all three transportation scenarios are shown in **Table 5**. The CI of the hydrogen produced is also presented with and without accounting for the downstream application of the waste biochar as a soil amendment.

For a better understanding of the breakdown of each factor contributing to the achieved CI for the hydrogen production process shown above, **Figure 7** demonstrates each factor separately for the three transportation scenarios.

Key findings from the analysis include:

1. The negative CI scores are due to the negative CI of the feedstock of biomethane (−269 g CO_2-e/MJ), as well as the CI credits generated due to the downstream application of the biochar to soil, leading to carbon soil sequestration. The CI of

Category	Name	Value	Units
Location	Madera, CA	—	—
	Fuel Pathway Inputs		
Feedstock	Biogas	13,187	MMBtu, LHV
Feedstock	Biochar	231	ton/month
Process energy	LH_2 Liquefaction	875,405	kWh
Transportation and Distribution	GH_2 Tube Trailer Truck	2,079,000	kg-miles
	LH_2 Tanker Truck		
	LH_2 Rail	21,637,000	
	GH_2 Fueling	9,083,888	MJ LH_2, LHV
	LH_2 Fueling	8,586,931	MJ LH_2, LHV
Co-product transport	Biochar transport	9,220,596	MJ biochar, LHV
Co-product agricultural effects	Biochar application to soil		
Co-product credit	Biochar soil C sequestration		

Table 4.
Key parameters for hydrogen production of the Hago energetics process using biogas as a feedstock. Biochar feedstock was from woody biomass from nearby forests and biochar was used as the catalyst for biomethane pyrolysis [32].

Scenarios	CI, $gCO_2e/MJ\ H_2$, LHV	CI, $kgCO_2e/kg\ H_2$
	CI with biochar carbon soil sequestration	
Scenario 1 (G.H2 by truck)[1]	−492	−59
Scenario 2 (L.H2 by truck)	−484	−58
Scenario 3 (L.H2 by rail)[2]	−484	−58
	CI without biochar carbon soil sequestration	
Scenario 1 (G.H2 by truck)[1]	−376	−45
Scenario 2 (L.H2 by truck)	−367	−44
Scenario 3 (L.H2 by rail)[2]	−371	−45

[1]*Transportation distance is 27 miles.*[2]*Transportation distance is 281 miles.*

Table 5.
CI score for hydrogen production with and without biochar carbon soil sequestration under the three transportation scenarios [32].

the biomethane and biochar production from pyrolysis was taken from the literature and the downstream transport of biochar was modeled in CA-GREET 4.0.

2. The results show that the positive transportation contributions to CI are very small compared to the large negative contributions of the biomethane feed and the biochar sequestration.

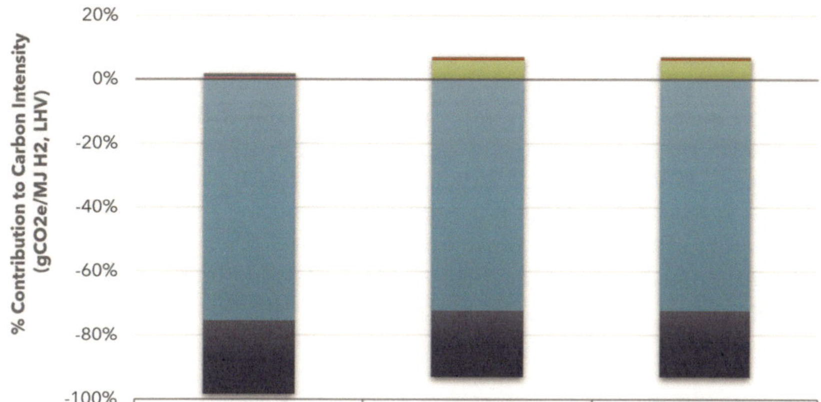

	Scenario 1 - GH2 (Truck)	Scenario 2 - LH2 (Truck)	Scenario 3 - LH2 (Rail)
Biochar soil C sequestration	-116.73	-116.73	-116.73
Biochar application to soil	0.25	0.25	0.25
Biochar transport	0.87	0.87	0.87
LH2 Fueling	0.00	4.03	4.03
GH2 Fueling	3.25	0.00	0.00
LH2 Rail	0.00	0.00	0.07
LH2 Tanker Truck	0.00	0.29	0.00
GH2 Tube Trailer Truck	2.75	0.00	0.00
LH2 Liquefaction	0.00	32.05	32.05
Feedstock - biochar	1.43	1.43	1.43
Feedstock - Biomethane	-383.72	-405.94	-405.94

Figure 7.
Contributions to carbon intensity for various scenarios [32].

BioMethod	Estimated Cost ($/kg H^2)	Smallest Carbon Footprint (kg CO^2eq/kg H^2)	Source
Gasification with CCS	10.3	−22.2	[34–37]
Biomethane Pyrolysis	10.5	−13.8	[34, 35, 38–41]
Biomethane SMR and ATR	8.0	−26.5	[36, 42–45]
Anaerobic Digestion	5.0	9	[34, 46–48]
Fermentation	8.6	0.96	[49–52]
Pyrolysis with Carbon Catalyst (Hago Method)	5.6	−59	[53]

Table 6.
Estimated Production Cost and Best Carbon Footprint for various biomass-based hydrogen production processes. After Ref. [33].

4.2 Summary of technoeconomic literature

A study from Columbia University [33] summarized the technoeconomic literature of biohydrogen production using different pathways. **Table 6** shows a comparison of their findings vs. the results shown above. The Hago process has the least cost and smallest footprint.

5. Conclusions

Biomass and biogas conversion to hydrogen technologies present an additional source for decarbonizing agricultural and industrial emissions. Conversion of biomass to hydrogen *via* thermochemical and biological processes has been described and comparisons have been made with these and conventional processes. Technoeconomic and LCA analyses have been illustrated including for a novel patented process under development. Research in unconventional technologies has been referenced. Some challenges of scalability and economics for obtaining hydrogen from biomass and reducing carbon emissions are mentioned.

Acknowledgements

We acknowledge support from the California Energy Commission through their CalSEED and CalTestBed programs for TEA and lab research support. We also acknowledge support from the US Department of Energy for their funding of third-party LCA estimates.

Nomenclature

CCSU	carbon capture sequestration and utilization
DMR	dry methane reforming
GHG	greenhouse gas
LCA	life cycle analysis
MP	methane pyrolysis
NZE	net zero emissions
SMR	steam methane reforming
TEA	techno-economic analysis

Author details

Wilson Hago[1*] and John O'Connell[2]

1 Hago Energetics Benefit Corporation, Camarillo, California, USA

2 O'Connell Associates, Nipomo, California, USA

*Address all correspondence to: wilson@hagoenergetics.com

References

[1] International Energy Agency (IEA). Global Hydrogen Review. 2024. Available from: https://iea.blob.core.windows.net/assets/3ece5ee4-7537-4992-8ba6-b83c994c3fd4/GlobalHydrogenReview2024.pdf [Accessed: October 17, 2024]

[2] Dincer I. Green methods for hydrogen production. International Journal of Hydrogen Energy. 2012;**37**:1954-1971. DOI: 10.1016/j.ijhydene.2011.03.173

[3] Cho H, Strezov V, Evans T. A review on global warming potential, challenges and opportunities of renewable hydrogen production technologies. Sustainable Materials and Technologies. 2023;**35**:e00567. DOI: 10.1016/j.susmat.2023.e00567

[4] Naqvi S, Kasmi B, Taqvi S, Chen W-H, Juchelková D. Techno economic analysis for advanced methods of green hydrogen production. Current Opinion in Green and Sustainable Chemistry. 2024;**48**:100939. DOI: 10.1016/j.cogsc.2024.100939

[5] Kamran M, Turzynski M. Exploring hydrogen energy systems: A comprehensive review of technologies, applications, prevailing trends, and associated challenges. Journal of Energy Storage. 2024;**96**:112601. DOI: 10.1016/j.est.2024.112601

[6] Lepage T, Kammoun M, Schmetz Q, Richel A. Biomass-to-hydrogen: A review of main routes production, processes evaluation and techno-economical assessment. Biomass and Bioenergy. 2021;**144**:105920. DOI: 10.1016/j.biombioe.2020.105920

[7] Rey J, Mateos-Pedrero C, Longo A, Rijo B, Brito P, Ferreira P, et al. Renewable hydrogen from biomass: Technological pathways and economic perspectives. Energies. 2024;**17**:3530. DOI: 10.3390/en17143530

[8] Singh S, Pandey G, Rath G, Veluswamy H, Molokitina N. Life cycle assessment of biomass-based hydrogen production technologies: A review. International Journal of Green Energy. 2023. DOI: 10.1080/15435075.2023.2245453

[9] Zhang L, Jia C, Bai F, Wang W, An S, Zhao K, et al. A comprehensive review of the promising clean energy carrier: Hydrogen production, transportation, storage, and utilization (HPTSU) technologies. Fuel. 2023;**355**:129455. DOI: 10.1016/j.fuel.2023.129455

[10] Tursi A, Olivito F. Biomass conversion: General information, chemistry, and processes. In: Mohammad Reza Rahimpour M, Kamali R, Makarem M, Manshadi M, editors. Advances in Bioenergy and Microfluidic Applications. Amsterdam: Elsevier; 2021. pp. 3-39. DOI: 10.1016/B978-0-12-821601-9.00001-7

[11] Damiri A, Stamatakis E, Bellas S, Zoulias M, Mitkidis G, Anastasiadis A, et al. A review of alternative processes for green hydrogen production focused on generating hydrogen from biomass. Hydrogen. 2024;**5**:163-184. DOI: 10.3390/hydrogen5020011

[12] Chandrasekhar K, Lee Y-J, Lee D-W. Biohydrogen production: Strategies to improve process efficiency through microbial routes. International Journal of Molecular Sciences. 2015;**16**:8266-8293. DOI: 10.3390/ijms16048266

[13] Noori M, Rossi R, Logan B, Min B. Hydrogen production in microbial

electrolysis cells with biocathodes. Trends in Biotechnology. 2024;**42**(7): 815-828

[14] AlHumaidan F, Absi Halabi M, Rana M, Vinoba M. Blue hydrogen: Current status and future technologies. Energy Conversion and Management. 2023;**283**: 116840. DOI: 10.1016/j.enconman.2023.116840

[15] Ananthi V, Bora A, Ramesh U, Yuvakkumar R, Raja K, Ponnuchamy K, et al. A review on the technologies for sustainable biohydrogen production. Process Safety and Environmental Protection. 2024;**186**:944-956. DOI: 10.1016/j.tibtech.2023.12.010

[16] Wang Y, Wu J. Thermochemical conversion of biomass: Potential future prospects. Renewable and Sustainable Energy Reviews. 2023;**187**:113754. DOI: 10.1016/j.rser.2023.113754

[17] Wu N, Lan K, Yao Y. An integrated techno-economic and environmental assessment for carbon capture in hydrogen production by biomass gasification. Resources, Conservation and Recycling. 2023;**188**(1):106693. DOI: 10.1016/j.resconrec.2022.106693

[18] Boretti A, Banik B. Advances in hydrogen production from natural gas reforming. Advanced Energy & Sustainability Research. 2021;**2**(11): 2100097. DOI: 10.1002/aesr.202100097

[19] Obiora N, Ujah C, Asadu C, Kolawole F, Ekwueme B. Production of hydrogen energy from biomass: Prospects and challenges. Green Technologies and Sustainability. 2024;**2**: 100100. DOI: 10.1016/j.grets.2024.100100

[20] Elhambakhsh A, Long N, Lamichhane P, Hessel V. Recent progress and future directions in plasma-assisted biomass conversion to hydrogen. Renewable Energy. 2023;**218**:119307. DOI: 10.1016/j.renene.2023.119307

[21] Martins A, Rouboa A, Monteiro E. On the green hydrogen production through gasification processes: A techno-economic approach. Journal of Cleaner Production. 2023;**383**:135476. DOI: 10.1016/j.jclepro.2022.135476

[22] Taqvi S, Kazmi B, Naqvi S, Juchelková D, Bokhari A. State-of-the-art review of biomass gasification: Raw to energy generation. ChemBioEng Reviews. 2024;**11**(4): e202400003. DOI: 10.1002/cben.202400003

[23] Dufour A, Celzard A, Fierro V, Broust F, Coursond C, Zoulaliane A, et al. Catalytic conversion of methane over a biomass char for hydrogen production: Deactivation and regeneration by steam gasification. Applied Catalysis A: General. 2015;**490**:170-180. DOI: 10.1016/j.apcata.2014.10.038

[24] Hamdani I, Ahmad A, Chulliyil H, Srinivasakannan C, Shoaibi A, Hossain. Thermocatalytic decomposition of methane: A review on carbon-based catalysts. ACS Omega. 2023;**8**(32): 28945-28967. DOI: 10.1021/acs.iecr.3c03895; 10.1021/acsomega.3c01936

[25] Hantoko D, Khan W, Osman A, Nasr M, Rashwan A, Gambo Y, et al. Carbon–neutral hydrogen production by catalytic methane decomposition: A review. Environmental Chemistry Letters. 2024; **22**:1623-1663

[26] US Patents 10,589,999 and 12,128,355

[27] Seider W, Lewin D, Seader J, Widagdo S, Gani R, Ng K. Product and Process Design Principles: Synthesis,

Analysis and Evaluation. 4th ed. New York: Wiley; 2017. p. 738

[28] Sphera. GaBi 4 Software [Internet]. Available from: https://sphera.com/solutions/product-stewardship/life-cycle-assessment-software-and-data/ [Accessed: October 10, 2024]

[29] SimaPro. Software [Internet]. Available from https://simapro.com/ [Accessed: October 20, 2024]

[30] Department of Energy. Software [Internet]. Available from: https://www.energy.gov/eere/greet [Accessed: October 24, 2024]

[31] Hijazi O, Munro S, Zerhusen B, Effenberger M. Review of life cycle assessment for biogas production in Europe. Renewable and Sustainable Energy Reviews. 2016;**54**:1291-1300. DOI: 10.1016/j.rser.2015.10.013

[32] Eco Engineers Report to Hago Energetics Benefit Corporation, Preliminary Carbon Intensity Calculations for Hydrogen Production, November 10, 2024.

[33] Lou Y, Fan Z, Friedmann J, Corbeau A-S, Agrawal M, Khatri A. The Potential of Biohydrogen in Creating a Net-Zero World: The Production and Applications of Carbon-Negative Hydrogen. New York; Columbia Center on Energy Policy; 2023

[34] Khan M, Ngo H, Guo W, Liu Y, et al. Biohydrogen production from anaerobic digestion and its potential as renewable energy. Renewable Energy. 2018;**129**: 754-768. DOI: 10.1016/j.renene.2017.04.02

[35] Kumar G, Eswari A, Kavitha S, Kumar MD. Thermochemical conversion routes of hydrogen production from organic biomass: Processes, challenges and limitations. Biomass Conversion and Biorefinery. 2020;**11**:1-26. DOI: 10.1007/s13399-020-01127-9

[36] Rosa L, Mazzotti M. Potential for hydrogen production from sustainable biomass with carbon capture and storage. Renewable and Sustainable Energy Reviews. 2022;**157**:112123. DOI: 10.1016/j.rser.2022.112123

[37] NREL. Hydrogen production cost estimate using biomass gasification: Independent review. National Renewable Energy Laboratory. 2011: NREL/BK-6A10-51726. Available from: https://www.hydrogen.energy.gov/pdfs/51726.pdf

[38] Liu L, Jiang P, Qian H, Mu L, et al. CO_2-negative biomass conversion: An economic route with Co-production of green hydrogen and highly porous carbon. Applied Energy. 2022;**2022**(311): 118685. DOI: 10.1016/j.apenergy.2022.118685

[39] Lu Q, Li W, Zhang X, Liu Z, et al. Experimental study on catalytic pyrolysis of biomass over a Ni/Ca-promoted Fe catalyst. Fuel. 2020;**263**: 116690. DOI: 10.1016/j.fuel.2019.116690

[40] Nasir Uddin M, Daud W, Abbas H. Potential hydrogen and non-condensable gases production from biomass pyrolysis: Insights into the process variables. Renewable and Sustainable Energy Reviews. 2013;**27**:204-224. DOI: 10.1016/j.rser.2013.06.031

[41] Sarkar S, Kumar A. Large-scale biohydrogen production from bio-oil. Bioresource Technology. 2010;**101**(19): 7350-7361. DOI: 10.1016/j.biortech.2010.04.038

[42] Milbrandt A, Bush B, Melaina M. Biogas and hydrogen systems market assessment. National Renewable Energy

Laboratory (NREL). 2016. Available from: https://www.nrel.gov/docs/fy16osti/63596.pdf

[43] Marcoberardino G, Vitali D, Spinelli F, Binotti M, Manzolini G. Green hydrogen production from raw biogas: A techno-economic investigation of conventional processes using pressure swing adsorption unit. PRO. 2018, 2018; **6**(3):19. DOI: 10.3390/pr6030019

[44] Minutillo M, Perna A, Sorce A. Green hydrogen production plants via biogas steam and autothermal reforming processes: Energy and exergy analyses. Applied Energy. 2020;**277**:115452. DOI: 10.1016/j.apenergy.2020.115452

[45] Zhou Y, Swidler D, Searle S, Baldino C. Life-Cycle Greenhouse Gas Emissions of Biomethane and Hydrogen Pathways in the European Union. The International Council on Clean Transportation; 2021. Available from: https://theicct.org/wp-content/uploads/2021/10/LCA-gasEU-white-paper-A4-v5.pdf

[46] Hajizadeh A, Mohamadi-Baghmolaei AM, Saady NC, Zendehboudi S. Hydrogen production from biomass through integration of anaerobic digestion and biogas dry reforming. Applied Energy. 2017;**309**:118442. DOI: 10.1016/j.apenergy.2021.118442

[47] Park M, Jo J, Park D, Lee D, Park J. Comprehensive study on a two-stage anaerobic digestion process for the sequential production of hydrogen and methane from cost-effective molasses. International Journal of Hydrogen Energy. 2010;**35**(12):6194-6202. DOI: 10.1016/j.ijhydene.2010.03.135

[48] Saidi R, Liebgott P, Gannoun, Gaida LB, et al. Biohydrogen production from Hyperthermophilic anaerobic digestion of fruit and vegetable wastes in seawater: Simplification of the culture medium of Thermotoga Maritima. Waste Management. 2018;**2018**(71):474-484. DOI: 10.1016/j.wasman.2017.09.042

[49] Djomo S, Humbert S, Blumberga D. Life cycle assessment of hydrogen produced from potato steam peels. International Journal of Hydrogen Energy. 2008;**33**(12):3067-3072. DOI: 10.1016/j.ijhydene.2008.02.006

[50] Ferreira A, Marques A, Batista A, Marques P, et al. Biological hydrogen production by anabaena sp.—Yield, energy and CO_2 analysis including fermentative biomass recovery. International Journal of Hydrogen Energy. 2019;**37**(1):179-190. DOI: 10.1016/j.ijhydene.2011.09.056

[51] Oncel S. Biohydrogen from microalgae, uniting energy, life, and green future. In: Kim SK, editor. Handbook of Marine Microalgae. Amsterdam: Elsevier; 2015. pp. 159-196. DOI: 10.1016/B978-0-12-800776-1.00011-X

[52] Osman A, Deka T, Baruah D, et al. Critical challenges in biohydrogen production processes from the organic feedstocks. Biomass Conversion and Biorefinery. 2023;**13**:8383-8401. DOI: 10.1007/s13399-020-00965-x

[53] Unpublished Data from Hago Energetics Benefit Corporation. 2022.

Chapter 6

Hydrogen Sulphide

Kofi Ofori

Abstract

Hydrogen sulphide (H_2S), a highly toxic and corrosive molecule, is typically found in hydrocarbon reservoirs, sewers and in the waste industry. It can be extremely problematic during drilling, production and processing. This chapter offers a synopsis of H_2S, which is sulphur in its most reduced form of all its numerous oxidation states. It delves briefly into H_2S's history on planet earth before there was life all through to its diminishment during the latter Proterozoic era to present day. It also investigates its various forms of generation and production, and its effect and impact especially as an occupation-based hazard. Its utilisation in enhanced oil recovery (EOR) as a standalone or together with carbon dioxide (CO_2) and its role in geosequestration together with CO_2 is explored.

Keywords: hydrogen sulphide, palaeohistory, geosequestration, generation, impact

1. Introduction

Hydrogen sulphide (**Figure 1**) is a colourless, rotten egg smelling gas that poses problems in many industries and occupations, especially in oil and gas and the waste industries due to the anaerobic nature of its production. Chemically, even though H_2S is a chalcogenide (a molecule composed of hydrogen and a group 16 element) just like water, it does not have the strong hydrogen bonds of water due to the oxygen atom in water's strong electronegativity, thus inducing the polarity in water that is missing in H_2S.

It usually occurs naturally in low-lying areas where there is a tendency for oxygen to be absent. The increase in exploitation of sour reservoirs [1] means that issues with H_2S will be ongoing, especially in deep carbonate reservoirs [2, 3] for the foreseeable future. It is toxic and corrosive and is immediately dangerous to life or health (IDLH) at just a 100 ppm. This therefore complicates any mix or environment H_2S is present in. Exposure to H_2S is largely occupational and not widespread, usually confined to energy sources from the subsurface such as oil and gas fields and others such as crude oil and natural gas refineries. Atmospheric H_2S usually is released through leaks from hydrocarbon field equipment including wellheads, sometimes through blowouts, and abandoned reservoirs. These reservoirs can contain sulphate-reducing bacteria (SRB), microorganisms (SRMs) or prokaryotes (SRPs). In the sewage industry, generation of H_2S could occur when the waste remains stationary for long periods of time, giving rise and conditions ripe enough for SRB, SRMs and SRPs to infest and become active. The primary function of the SRB, SRPs and SRMs is to break down (reduce) sulphur-bearing compounds into H_2S throughout the course of metabolism. Sulphur dioxide

Figure 1.
H_2S molecule with white representing the two hydrogen atoms and yellow sulphur.

(SO_2), the subject of this book, is related to H_2S as a product through a limited combustive oxidation reaction (and more importantly as the first reaction in the two-stage Claus desulphurisation process);

$$2H_2S_{(g)} + 3O_{2(g)} \rightarrow 2SO_{2(g)} + 2H_2O_{(l)} \quad (1)$$

2. A brief palaeohistory

Before the first life on earth developed in the late Eoarchaean-early Paleoarchaean eras nearly 4 billion years ago (4 Ga), H_2S was present, due to an abundance of sulphur, and anoxygenic photosynthesis was occurring between H_2S and CO_2 [4]. Although the process' time of evolution may be debated [5], oxygenic photosynthesis, aided by cyanobacteria also involving both acid gases, is believed to have commenced during the Archaean expansion of about 3.2–2.8 Ga ago (tail end of the Paleoarchaean all through to the very end of the Mesoarchaean era) [6], even though there was practically very little oxygen [7]. Some of these early life cyanobacteria, through their activities in the formation of stromatolites, can still be found in the Pilbara Craton's Strelley pool chert [8] and Warrawoona group of fossils [9], Cervantes' Lake Thetis [10] and Shark Bay [11], all in Western Australia. The great oxidation event (GOE), which occurred around the generally accepted ~2.4–2.1 Ga ago (middle Siderian through all the Rhyacian period) is hypothesised to have marked the enduring presence and spread of atmospheric oxygen from the oceans through cyanobacterial activity. These cyanobacteria used water as fuel to conduct photosynthesis by oxidising it into oxygen, and this produced oxygen filled out the atmosphere so fast in a ~ 300 million year period, threatening the very existence of the same oxygen-producing cyanobacteria. It also has been theorised that this event coincided with a substantial growth in oceanic sulphates [12]. But atmospheric anaerobiosis (life without free, molecular oxygen) shrunk and this affected the H_2S-producing cyanobacteria, driving them into low-lying spaces that were still anoxic. A focused study of the Proterozoic aeon indicates that in the periods that followed the GOE in the Paleoproterozoic era (Orosirian and Statherian), the oceans remained sulphidic, and by extension anoxic, due to production of sulphates from sulphur oxidation possibly aided by SRB, which was eventually swept out to the oceans [13]. The Mesoproterozoic era (starting 1.6 Ga ago) is remarkable for the increase in oceanic oxygen [14] and the development of multicellular eukaryotes over prokaryotes [15]. Note that prokaryotes are the basis for SRPs hence their diminution would mean a reduction in available sulphur and its other forms including H_2S. The rise of eukaryotes is also presented as evidence of the existence of atmospheric oxygen in this time period [16], even though the presence of H_2S in the atmosphere is proposed to have been responsible for the delay in eukaryotic development in this aeon [17]. Much of the atmospheric H_2S disappeared during the latter deglaciation years of the

Proterozoic aeon in the Neoproterozoic era, and specifically in the early parts of the Ediacaran period. This was after the waning thaw of the Cryogenian period after both the Sturtian and Marinoan snowball earth glaciation—between ~720 and 630 Ma ago [18]—when an influx of oxygen engulfed the atmosphere [4]. This is a little surprising considering even the GOE which essentially served as a bridge between the Archaean and Proterozoic eons and initiated multicellular life forms when free oxygen appeared in the atmosphere in voluminous amounts was not as successful in removing much of the H_2S present, and the GOE had a lot more oxygen. More perplexing still is that the GOE occurred simultaneously with the peak period in euxinic conditions, where oceans and water bodies had high H_2S content [19] and had little or no oxygen content except for at the surface [20]. In the current Phanerozoic eon, the Cambrian explosion of fossil biodiversity has been partly attributed to the cessation of euxinic oceanic conditions [21]. Mass extinctions earlier in in this eon at the tail end of the Permian period have been attributed in part to the accumulation of noxious H_2S in both the oceans and the atmosphere as well as CO_2-inducing hypercarbia [22]. An oxygen isotope study theorises that sulphates were probably lower in the early part of this Cenozoic period [23], hence lowered presence of H_2S to oxidate and/or not enough sulphates to reduce. Another study posits an oceanic sulphate fluctuation in the current period, with a 10 million year decrease starting 65 Ma ago, an increase for 10 Ma thereafter, then steady for the interval in between till the last 2 million years when it has been uniformly decreasing [24]. Present day, less than 1% of all ocean floors are covered by H_2S [25].

3. Generation

Microorganisms that reduce sulphates into H_2S (SRMs, SRPs, SRB) are found in every ecosystem on earth, from water bodies to soils and even in human and animal guts [26]. But most recent H_2S occurrence results from the production and processing of sour fluids in oil field operations. Sour fluids are defined in the oil and gas industry as any natural gas or petroleum fluids that contain more than 4 ppm of H_2S [27] at standard temperature and pressure (STP) conditions of 0°C (273.15 K) and 1 bar respectively. Whilst hydrogen sulphide (H_2S) formation/generation is complex and all processes involved are not entirely understood [28], it is known that H_2S can be formed either naturally or by industry production operations, especially during secondary and enhanced phases of recovery. A common denominator for each is the requirement of the absence of oxygen (anoxic or anaerobic conditions). As hydrocarbons also form in the absence of oxygen, then these hydrocarbon reservoirs serve as excellent media for *in situ* generation of H_2S. The process by which H_2S forms in reservoirs in the absence of oxygen, but in the presence of oxidising agents (oxidants) and/or nutrients together with hydrocarbons, is referred to as reservoir souring. Marriott et al. [29] explained three mechanisms for the generation of H_2S in the petroleum industries. These are through the breakdown of reservoir biomass or aquathermolysis, the breakdown and reduction of sulphate-bearing minerals through microbes, microbial sulphate reduction (MSR), and also through heat, or thermochemical sulphate reduction (TSR). They further stress that some operations in the oil and gas industry can affect H_2S production either by contributing towards its increased production or delaying its emergence. For instance, sea water injected into reservoirs to enhance recovery has also been shown to introduce microbes and initiate H_2S generation. Higher rates of H_2S generation result when cold sea water is pumped into oil zone but lower rates when cold water is injected near the water zone below oil zone [30]. This H_2S generation has been attributed to large quantities of sulphate being imparted by the sea water which is then worked on and reduced by SRPs in the reservoir [31]. These electron donor-oxidising prokaryotes, which take in sulphates and generate sulphides [32] are not restricted to just the reservoir but they can also sour production pipework and well tubing and offshore topside facilities [33]. The technologies used in unconventional oil and gas resources during the exploitation and development such as hydraulic fracturing and steam-assisted gravity drainage (SAGD) could possibly add more H_2S to the reservoir or turn existing sulphur in the reservoir to H_2S, Marriott et al's study appends [29]. Another research showed that the chemical reaction in crude petroleum between sulphoether and sulphur-alcohol may be responsible for the generation of H_2S, in addition to the thermochemical reactions of sulphate-containing minerals [34]. Anthropogenically-induced H_2S generation, especially in reservoirs that were initially sweet, usually occurs through production activities such as waterflooding. Ma et al. [35] conducted experiments based on superheat degree during thermal enhanced oil recovery (EOR) processes. They found that at a constant reservoir temperature, H_2S generation was more likely to occur at a greater superheat temperature, with a superheat temperature of 62.19°C recording H_2S formation of 0.178 mL/g whereas a superheat temperature of 89.42°C reported an H_2S generation of 0.345 mL/g. They also established an inverse relation between the increase in H_2S and the dwindling of sulphur-generating sources such as sulphides and sulphates, and typically non-sulphur generating sources such as methylene, carboxyl and carbonyl groups. H_2S is linked to CO_2 in that the first method of H_2S generation (through reservoir biomass) occurs when carbon-essential organic matter decomposes to form H_2S. This is often a by-product of the decomposition of the same organic matter to usually form aliphatic hydrocarbons (alkanes, alkenes and alkynes). The organic matter may include dead and rotting vegetation, coral reefs, plankton, and even animals. In fact, H_2S generation may even be in direct competition to hydrocarbon generation (a heterotrophic relationship) as methanogenesis (the generation of methane CH_4, the foundational aliphatic hydrocarbon) from peat bog sources may be delayed due to the effect of these sulphate-reducing microorganisms (SRMs) [36]. A sample study from the Santanghu Basin found that the ^{34}S isotopologue contributed up to a fifth of the sulphur in the areal H_2S and concluded that TSR and/or the thermal decomposition of sulphur (TDS) was/were the key generation mechanism(s) [37] and listed a TSR formation equation as thus [38]:

$$sulphate + petroleum \rightarrow calcite + H_2S \pm H_2O \pm CO_2 \pm S \pm altered\ petroleum \quad (2)$$

The methionine-to-serine amino acid metabolic transsulphuration pathway occurs when sulphur is transplanted to the amino acid l-cysteine from l-homocysteine and the process eventually culminates in the generation of H_2S [39]. Cysteines and the uptake of the sulphur-containing cystine are particularly important in oncogenesis, when previously healthy cells start becoming carcinogenic [40]. The sulphur in persulphides (RSSH) can also be reduced to the endogenous signalling biomolecule H_2S, which plays a paramount role in some mammalian physiology [41]. A simplified reaction is as follows:

$$R - SSH + R - SH \rightarrow R - SS - R + H_2S \quad (3)$$

Simply put, a cysteine persulphilde is reduced by the organosulphur thiol to produce the disulphide cystine and H_2S [42] where R is a functional alkyl or organic substituent.

The presence of humic acid in dissolved organic matter (DOM) in sediments and soils could also serve as a key driver in H_2S generation [43] and can also be responsible for the multiplication of microorganisms in sediments and soils [44] which is favourable for both H_2S and hydrocarbon generation. Humic acid can also hasten or hinder hydrocarbon generation through control over biodegradation solubilisation and limiting cell adhesion [45] but petroleum hydrocarbons (PHs)—made up of straight chain aliphatic hydrocarbons (AHs) and polycyclic aromatic hydrocarbons (PAHs)—can also have their phytodegradation amplified by the humic acid [46]. In soils and sediments, humic matter—humic acids, fulvic acids and humin—are classified as the key ingredients in organic composition when animals and plants wither and decay through the biological and chemical agents and actions of microorganisms [47]. In the sulphur cycle, fulvic acids may also act as sinks (geological storage or reservoir) in deep groundwater for when H_2S deliquesces in water [48].

Generation of H_2S in other industries is also well documented though in scale and size, is dwarfed by generation in hydrocarbon reservoirs. In the waste and sewage industry, Zhiqiang et al. [49] relate factors that affect H_2S generation including isolating the manhole septic tank as a major hotspot, the effects of temperature and flow turbulence, water quality and how heavy rains hinder H_2S production because of sheer moving velocity. In thermal springs or geothermal fields, sulphur-bearing minerals deep in the earth oxidise to form H_2S which in turn dissolves in the hot water. As this hot water rises to the surface and the water gets colder, the H_2S evolves from the water giving the spring the rotten egg smell. These types of springs are called sulphurous thermal waters (STW). Industries as diverse as tanneries, piggeries, pulp and paper manufacturing, metallurgical and thermal coal plants, textile production especially rayon, landfills and manure storage pits have also been known to produce H_2S. With coal, H_2S is usually generated or 'sorbed' out during high-pressure but below calcination temperature gasification by limestone in the form of calcium carbonate [50] whilst SO_x is a/are direct product(s) when combustion takes place instead [51]. Biomass—in the form of biofuels, bioliquids, biogas, etc.—a source of renewable energy touted as one of the potential fossil fuel substitutes, also releases amounts of H_2S during the processing into energy form [52]. Whilst most the generation is through SRMs and chemolithotrophs (organisms that derive chemical energy and sustenance through the oxidation of inorganic sources such as rocks), the influence of temperature (thermochemical and thermobiological) is also felt as even under anaerobiosis, the sulphate-reducing bacteria (SRB) thrive under warm conditions. Underwater thermal vents, volcanoes, stagnant waters and polluted waters can also produce H_2S [53]. Stoichiometry-wise; one of the primary SRB reactions to produce H_2S is:

$$SO_4^{\ 2-} + 10H^- + 8e^- \overset{SRB}{\rightarrow} H_2S + 4H_2O \tag{4}$$

And two others that convert the intermediate monosaccharide metabolite glyceraldehyde;

$$2SO_2 + C_3H_6O_3 \overset{SRB}{\rightarrow} 2H_2S + 3CO_2 + H_2O \tag{5}$$

$$3SO_4^{\ 2-} + 2C_3H_6O_3 \overset{SRB}{\rightarrow} 3H_2S + 6HCO_3^{\ -} \tag{6}$$

In wastewater systems, abundance of sulphate ions results in their reduction to an intermediate sulphur ion and then conversion to H_2S thus;

$$SO_4{}^{2-} + organic\ matter \overset{SRB}{\rightarrow} S^{2-} + H_2O + CO_2 \tag{7}$$

$$S^{2-} + 2H^+ \overset{SRB}{\rightarrow} H_2S \tag{8}$$

All the above-listed equations occur anaerobiotically, that is, the microorganisms perform them without the benefit of free, molecular oxygen. Outside of SRB equations, hydrogen gas and elemental sulphur can also react to form H_2S;

$$H_{2(g)} + S_{(s)} \rightarrow H_2S_{(g)} \tag{9}$$

The reaction between aluminium sulphide and water also achieves that purpose;

$$Al_2S_{3(s)} + 6H_2O_{(l)} \rightarrow 2Al(OH)_{3(s)} + 3H_2S_{(g)} \tag{10}$$

And another equation between ferrous sulphide and hydrochloric acid usually in a Kipp's apparatus in a fume hood also produces H_2S;

$$FeS_{(s)} + 2HCl_{(aq)} \rightarrow FeCl_{2(s)} + H_2S_{(g)} \tag{11}$$

Solid thioacetamide decomposes in hot water to form acetate and ammonium ions (acetamide or ammonium acetate) and is most importantly a precursor for H_2S production;

$$CH_3CSNH_{2(s)} + 2H_2O_{(l)} \overset{heat}{\rightarrow} CH_3COO^-{}_{(aq)} + NH_4^+{}_{(aq)} + H_2S_{(g)} \tag{12}$$

The above listed are by no means exhaustive. H_2S can be produced several different ways synthetically in the lab but these are the major means by which H_2S is generated. Approximately 90% of H_2S found in the environment occurs naturally [54].

4. Effects and impact

In the petroleum field, H_2S can make oil and gas reservoirs harder to develop and can hinder hydrocarbon extraction from formation reservoirs as they are classified as contaminants in natural gas which consists of mainly methane and ethane and thus the natural gas has to be "sweetened" [55] to remove these impurities. Oil and gas workers may also be exposed to the potentially fatal H_2S that may be present in drilling fluids as the gas may be brought up to the surface when drilling mud goes through its cycle. H_2S is corrosive and can make both concrete and steel equipment in the upstream flowlines and tubing brittle [56]. It also has a similar corrosive effect on iron, copper and brass [57], and its presence leads to increased corrosion rates [56].

In waste water and sewerage treatment plants, biogas plants, industrial and manufacturing processes, chemical, petrochemical and oil refining processes, during production, storage and transportation, H_2S is known to be highly toxic even at low concentrations. It has been proven to cause cognitive impairment [58] and has also been linked to neurological symptoms and health effects on the central nervous system and respiratory functions [59]. In a study of nearly 150 wastewater workers, for whom H_2S is frequently an occupational health and safety hazard, an inflammation was found, albeit low but pervasive, symptomatic of the effect on lung capacity through exposure to workaday strength H_2S [60]. H_2S is usually the unwanted end-product of the hydrodesulphurisation of coal and hydrocarbons and may interrupt normal physiological functions and cause ailments such as asthma, sepsis, neurodegenerative disease, cirrhosis and diabetes [54]. Elevated concentrations of H_2S in the gut can be detrimental to the intestinal and bowel regions because it reduces pH and can induce toxicity [61].

Studies have shown that some prokaryotes (primarily archaea) and bacteria utilise sulphur compounds through their oxidation or reduction to aid and invigorate both cellular and chemolithotrophic growth [62].

5. Mitigation

Large-scale H_2S produced from hydrocarbon reservoirs usually undergo a hydrodesulphurisation Claus process where in the presence of a catalyst it is converted into elemental sulphur. This has been the established choice for at least the last few decades. Because H_2S produced in the waste and sewage is localised in nature, it is easier to fashion containment strategies, especially through chemical dosing using nitrates and biocides [63] for prevention and remediation purposes. But these biocides may require more space and more importantly their effluvia can possibly be carcinogenic hence a nitrate-based solution whereby SRB are inhibited by the growth and enrichment of the less-unpleasant nitrate-reducing bacteria (NRB) [33]. A free ammonia based method wherein urine is dosed in the sewage system was found to greatly reduce H_2S production by the genus *Desulfobulbus* [64]. A free nitrous acid (FNA) dosing was also observed to substantially lower operational costs and leave a much lower carbon footprint [65]. The growth of the twin sulphide oxidisers *Rhizobiaceae* and *Xanthomonadaceae* was reinvigorated by nitrates [66]. A study [67] found that nitrate-reducing, sulphide-oxidising bacterium (NR-SOB) consisting of the culture *Thiomicrospira sp. strain CVO* dosed in an SRB environment rife with *Desulfovibrio sp. strain Lac6 and t*he NR-SOB obstructed the growth of the SRB. In the presence of high level of the NR-SOB, complete oxidation of the sulphide took place, totally eliminating any remaining H_2S, the study adds [67]. The nitrate-reducing bacteria work by competing directly with SRB and heterotrophically beating it for available organic matter to biodegrade [31]. They also oxidise H_2S already produced [68]. Sulphidogenesis, the process of sulphide generation, has also been controlled with (per)chlorates by energising two known chlorate respirers *Pseudoalteromonadaceae* and *Pseudomonadaceae* through biologically competitive exclusionary means [66]. Other measures of control include the use of bacteriophages, inorganic chemicals, sonication, filtration and disinfecting with UV rays [69]. Scavengers such as zinc oxide (ZnO), copper(II) nitrate trihydrate ($Cu(NO_3)_2{\cdot}3H_2O$) and potassium permanganate ($KMnO_4$) of varying mass concentrations in drilling mud have been used historically to reduce the threat posed by H_2S in drilling fluids when they are cycled during oil and gas drilling operations [70]. Further, it has been established that in sewer pipes, the combination of conductive concrete containing magnetite, and the presence of electricity-producing bacteria reduces the concentration of H_2S [71].

Together, however, and in much larger volumes than found at waste sites, the hydrocarbon reservoir-produced acid gases (primarily CO_2 and H_2S) can be used in Enhanced Oil Recovery (EOR) [72]. This is whereby they are reinjected into formations to usually boost tertiary production of hydrocarbons after the natural pressures in the reservoir significantly reduce over time during primary production. This specific EOR, known as flue gas flooding, is on the ascendancy because components such as CO_2, H_2S, CO, N_2, NO_x and SO_2 are all part of the mix that is injected. These flooding operations contribute towards reducing greenhouse emissions. The presence of H_2S is known to increase the sweep efficiency of the residual oil due to its greater density and being more soluble in the residual oil compared to other acid gases [73]. The rise in acid gas flooding is also attributable to its inexpensive nature and the ease of access [74]. The acid gas mix can also be used as part of a combination flooding strategy (together with water flooding and surfactants) where their combined effects may serve to improve the displacement efficiency of the unswept oil region [75]. Bolstering the case even further for their joint use is research that shows that an acid gas mixture containing over four-fifths CO_2 concentration—which is usually the case except in very sour reservoirs—is almost azeotropic in nature [76]. This makes it difficult to separate the mixture through binary distillation and the alternative process may be energy-intensive and solvents may be required to proceed with separation (for which m-xylene was discovered to be the best amongst four studied) [76]. This phenomenon is analogous to why ethanol cannot be absolutely distilled to 100% purity because of the non-ideal azeotropic behaviour; at a certain ethanol concentration—usually just under a 100%—the vapour phase concentration of both alcohol and water becomes equivalent to the liquid phase's concentration of both, hence any extra distillation produces both water and ethanol and not the required ethanol, resulting merely in a waste of energy.

6. Role in geosequestration

Utilising or geosequestering both acid gases together make sense in many ways especially if they are produced together. Economically, disposing of both together saves costs compared with storing them individually. They can be both used in enhanced oil recovery (EOR) operations and their injection could be used to maintain reservoir pressures [77]. Additionally, and perhaps significantly, research has shown that the EOR injection of the acid gas mixture has a delayed breakthrough at the producing well compared to a CO_2 EOR injection alone [78]. Furthermore, there can be another case made for the sequestration of these gases especially for H_2S, as the alternative, the desulphurisation process that ends in the reduction of the produced H_2S into sulphur [79] is proving to be financially unfeasible [80] mainly due to a reduced global appetite for elemental sulphur.

In CO_2 and H_2S sequestration and storage projects, an important concern is to ensure that the gases will be securely contained for the long term in the given geological formation. According to International Energy Agency Greenhouse Gas Research and Development Programme (IEAGHG), 'long term' is classified as a period anywhere from several hundred to thousands of years [81].

From a storage efficiency (minimum leakage) perspective, reservoirs (storage sites) may be considered based on their ability to trap the fluids injected, and the formation fluids (oil and water) will be more dense than the injected fluids [82]. Theoretically and practically, injected fluids are prevented from escaping to the surface by four trapping mechanisms: structural trapping, residual trapping, dissolution trapping and mineral trapping [83].

For multiple phase conditions, surface wetting in a hydrocarbon reservoir, analyses of the behaviour of the supercritical acid gas, and its eventual fate in the reservoir after sequestration, requires an in–depth knowledge of the processes, molecular reactions and interactions between mineral and fluid phases to create larger–scale simulation models [84]. For a safe and efficient CO_2, H_2S or acid gas storage project, the accurate prediction of reservoir molecular interactions is very important.

In 1989, Chevron injected acid gas containing 85% CO_2 by volume and 15% H_2S in Acheson, Alberta, Canada [85] in what is perhaps the first of its kind. Whilst acid gas injection (AGI) operations may not be as common as carbon capture and storage (CCS), many projects have been investigated, with over 80 acid gas sequestration operations in the world over a decade ago, especially in Canada [86], Poland [87] and Algeria [88]. The geosequestration processing of acid gases is generally classified into two main actions: the surface operational exercises which involves the capture, transport and injection of the gas, and the *in situ* processes which relate to the storage functions within the geological media [89]. Large-scale H_2S capture is usually from sour oil and gas reservoirs unlike CO_2 which in addition to being captured from hydrocarbon reservoirs can also be directly captured from the atmosphere.

In the oil and gas industry, produced fluids containing significant amounts of H_2S and CO_2 are typically passed through a processing and refinery plant that contains two essential components: an absorption column and a regeneration tower. The former, also called an amine gas treatment column, usually uses an aqueous alkanolamine variant such as;

- diethanolamine (DEA),
- triethanolamine (TEA),
- diglycolamine (DGA),
- diisopropanolamine (DIPA),
- monoethanolamine (MEA),
- methyldiethanolamine (MDEA),
- aminomethyl propanol (AMP),

Sometimes a hybrid mixture is used to absorb the acid gases out of the produced gas stream, hence generating a hydrocarbon phase devoid of or containing low amounts of acid gas (sweet gas or crude). The stream containing the acid gas and the amine is then passed through the regeneration column where the acid gas is separated from the amine and the amine can be reused in the absorption column for future separation.

Separated acid gases are usually compressed and temporarily stored, transported and injected in a supercritical state where they are neither liquids nor gases but enjoy the distinct benefit of being very diffusive, like gases, but nearly as dense as liquids.

For simplicity and economic reasons, wellhead injection temperatures and pressures are usually aligned with that of the compressed gas's transportation conditions [90] although temperatures can range between 30°C all through to 200°C and pressures from 1 to 30 MPa, with both temperature and pressure ceiling usually determined by the reservoir's state [91], and the acid gas injection strategy delineated to take advantage of system thermodynamics, including importantly, use of the phase envelope [92]. Storage is usually implemented in deep saline aquifers, depleted oil and gas reservoirs, clastic and basaltic formations, unmineable coal seams [93], and unconventional oil and gas reservoirs [94]. The temperatures are non-trivial during injection, as they affect the reservoir fluids and rock thermodynamically, chemically, hydrologically and mechanically, wherein the Joule-Thomson cooling effect may occur due to advection and expansion of injected fluids [90]. This disruption to a system previously in equilibrium can lead to continued effects on the reservoir and cap rock as it may cause stress/strain on them, opening up old fissures and faults or creating new fractures hence affecting long term containment security and safe storage [95].

Factors that go into selecting a geological disposal media are broadly framed under physical, hydrodynamic, regulatory, geological, social, thermodynamic, economic and technical [96]. They include the minimisation of geomechanical risks, site selection, retention and optimising injectivity [97] containment security and reservoir capacity. An expansive and exhaustive list laid out by Raza et al. [98] include common benchmarks such as permeability, density, depth, porosity, trapping mechanisms, injectivity, minerals present, residual hydrocarbon saturation, rock type, sealing potential, well type, wettability, subsurface type, etc. Studies have shown that deep saline aquifers have larger storage capacities compared to depleted oil and gas reservoirs [99] with the former having a range anywhere between 1000 and 10,000 Gt compared to a maximum of 900 Gt of storage [100] as original reservoir pressures during oil/gas producing life should not be superseded lest overlying cap rocks may be impaired and may give rise to leakages and storage integrity compromised [101]. Economically, a low capital outlay on operational frameworks (infrastructure) and an ability to sequester buoyant fluids safely in geological media and time are some of the reasons that make depleted oil and gas reservoirs appealing [99, 102]. Added to that is the fact that they have two other advantages: they once were producing reservoirs so they'll have been studied immensely through their geology, core and well logs and production logs so they will have rich historical data, and secondly, because they have a history of trapping oil and gas prior, they therefore make good candidates to trap fluids being injected [103]. Depleted gas reservoirs are also understood to be a more secure outlet for injection than other geological receptacles [102] since saline aquifers may be subject to a buildup of pressure over a region due to water's mobility [104].

There is more risk attached to the *in situ* geosequestration phase as it comes with a greater degree of uncertainty [89], which is why a rigorous study of the interface (interfacial tension, thickness, etc) is important to reduce the risks involved in the process. This is especially imperative considering the recent problems both the seminal Sleipner and Snøhvit CO_2 injection operations in Norway have run into. Sleipner has had CO_2 migrate from the formation fluid and 'hover' just underneath the caprock whilst Snøhvit has had its original 18 year storage capacity reduced to less than 2 years [105]. The Borzęcin sequestration project in Poland is unique in that it injects the acid gas into the water zone overlain by a producing gas cap and not as an enhanced gas recovery (EGR) technique but as an ongoing study of the effects of the acid gas on the formation water, and once formation water reaches critical saturation and the injected gas starts buoyantly moving upwards, the effects on the producing gas cap [106]. Some of the injected acid gas will be soluble in the formation water. In that case what will be the implications for newly formed weak acidic fluid, the interactions within the new fluid and its relationship with the reservoir rock? A weak acid will still lower the pH of the formation fluid and thereby increase solubility of the injected fluid. Both acid gases are diprotic, that is, they undergo two ionisation reactions when they dissociate to release H^+ or H_3O^+ ions, with H_2S shown thus;

$$H_2S_{(aq)} \rightleftharpoons H^+{}_{(aq)} + HS^-{}_{(aq)} \tag{13}$$

$$HS^-{}_{(aq)} \rightleftharpoons H^+{}_{(aq)} + S^{2-}{}_{(aq)} \tag{14}$$

CO_2 in water presents as weak carbonic acid, which is very important in the carbon cycle as it is responsible for the storage of CO_2 in carbonate rocks during weathering [107], and the two dissociation steps written thus;

$$H_2CO_{3(aq)} \rightleftharpoons H^+{}_{(aq)} + HCO_3{}^-{}_{(aq)} \tag{15}$$

$$HCO_3{}^-{}_{(aq)} \rightleftharpoons H^+{}_{(aq)} + CO_3{}^{2-}{}_{(aq)} \tag{16}$$

However, the two ionisation reactions for both CO_2 and H_2S typically have very low equilibrium constants, *K*. Since *K* is several orders lower than 1 (order usually between 10^{-7} and 10^{-19}), then both acid gases will exist in the formation water in their molecular form (CO_2 and H_2S) compared to in their dissociative states ($HCO_3{}^-$, $CO_3{}^{2-}$ and HS^-, S^{2-} respectively) as equilibrium would overwhelmingly favour the reactants. This very likely will be temporary until geochemical reactions between injected fluids, *in situ* fluids and reservoir rock commence. Some research suggests the new fluid will dissolve some of the reservoir rock minerals which in turn may create secondary reactions and new products hence the need to study thermodynamic, geochemical, structural and physical properties [108]. A lot of research has been done, especially for deep saline aquifers and the fluid-fluid-rock environment [109–112]. Our group's work on both injected H_2S [113] and acid gas (CO_2/H_2S) [114] using molecular dynamics (MD) simulation showed that geosequestration in depleted oil and gas reservoirs is ideal at low pressures and higher temperatures as more of the injected fluid is absorbed into the formation water under these conditions. Higher absorption rates reduce the risk of the injected fluid buoyantly escaping the subsurface environment. However, these conditions are also very germane for the generation of clathrate hydrates, with H_2S in particular singled out as the most amenable 'guest' molecule for hydrate formation [92], so a balance needs to be struck.

Research on H_2S-water and acid gas-water systems lags behind CO_2-water due to the difficult reservoir conditions and the toxic and corrosive nature of H_2S. Most research has been on enhanced oil recovery (EOR) [72, 115]. Bachu and co-workers [116–118] have done extensive work in Canada on CO_2 and acid gases during geosequestration using analytical models to approximate behaviour of injection plumes. Isothermal cracking experiments have also been performed on heavy oil samples to resolve the H_2S generation mechanism and it demonstrated that increasing the cracking temperature invariably increased the H_2S yield [119]. H_2S has also been injected in small rates in hydrothermal reservoirs in Iceland as a pilot, with prospects for later expansion [120]. Metal-organic frameworks (MOFs) have been recognised for their adsorbency in gas separation and storage due to their high selectivity of the corrosive H_2S [121, 122]. Koschel and co-workers [123, 124] have also performed considerable research and reviewed enthalpy and solubility data for H_2S under geological storage conditions. Chapoy et al. [125] developed a vapour-liquid equilibrium (VLE) model for the H_2S-water system at conditions (temperatures between 25 and 65°C, and pressures up to 4 MPa) below typical reservoir conditions. Possibly the widest range of temperatures and pressures parametrised for the H_2S-water binary system and the acid gas-water ternary system was by Springer et al. [126] who developed VLEs for these systems by reviewing several different data points in the literature. A considerable number of diagenetic, hydrogeological, mineralogical and stratigraphic appraisals have been performed [127] for acid gas injection and geosequestration. In all these works however, they have not *necessarily* looked at the intermolecular interactions, but at more specific properties and macroscopic scale behaviours. The work of Ng et al's [128], stressed the need for proper property calculations, especially density, near the critical point of the acid gas mixture as this is at injection conditions and properties are functionally very dependent on pressure and temperature.

7. Conclusion

Hydrogen sulphide (H_2S) is a molecule that presents mostly in oil and gas reservoirs and processing operations. Due to its acutely toxic and corrosive nature, it represents a specific complication during operations. This chapter attempts to highlight some of the issues. It also briefly traced its presence on earth before the genesis of life in the Archaean all through to its role in different timelines to the present Cenozoic era. Generation and production mechanisms have also been discussed, with emphasis on natural production which accounts for the majority of H_2S on the planet. Its effects as mainly an occupational health and safety risk, together with mitigation strategies are also deliberated upon. Finally, its topical and relevant position as a constituent acid gas injected for geosequestration is also briefly examined.

Conflict of interest

The author declares no conflict of interest.

Author details

Kofi Ofori
Curtin University, Perth, Australia

*Address all correspondence to: kofi.ofori@postgrad.curtin.edu.au

References

[1] Zhang X, Li Q, Zheng L, Li X, Xu L. Numerical simulation and feasibility assessment of acid gas injection in a carbonate formation of the Tarim Basin, China. Oil & Gas Science and Technology - Revue d'IFP Energies nouvelles. 2020;**75**:28

[2] Jiao Y, An L, Wang W, Ma J, Wu C, Wu X. Microbial communities and their roles in the Cenozoic sulfurous oil reservoirs in the southwestern Qaidam Basin, western China. Scientific Reports. 2023;**13**(1):7988

[3] Nöth S. High H_2S contents and other effects of thermochemical sulfate reduction in deeply buried carbonate reservoirs: A review. Geologische Rundschau. 1997;**86**(2):275-287

[4] Olson KR, Straub KD. The role of hydrogen sulfide in evolution and the evolution of hydrogen sulfide in metabolism and signaling. Physiology. 2016;**31**(1):60-72

[5] Cardona T, Murray JW, Rutherford AW. Origin and evolution of water oxidation before the last common ancestor of the cyanobacteria. Molecular Biology and Evolution. 2015;**32**(5): 1310-1328

[6] Lepot K. Signatures of early microbial life from the Archean (4 to 2.5 Ga) eon. Earth-Science Reviews. 2020;**209**: 103296

[7] Catling DC, Zahnle KJ. The Archean atmosphere. Science Advances. 2020; **6**(9):eaax1420

[8] Allwood AC, Walter MR, Burch IW, Kamber BS. 3.43 billion-year-old stromatolite reef from the Pilbara Craton of western Australia: Ecosystem-scale insights to early life on earth. Precambrian Research. 2007;**158**(3): 198-227

[9] Derenne S, Robert F, Skrzypczak-Bonduelle A, Gourier D, Binet L, Rouzaud J-N. Molecular evidence for life in the 3.5 billion year old Warrawoona chert. Earth and Planetary Science Letters. 2008;**272**(1):476-480

[10] Logan BW. Cryptozoon and associate stromatolites from the recent, Shark Bay, western Australia. The Journal of Geology. 1961;**69**(5):517-533

[11] Wacey D, Urosevic L, Saunders M, George AD. Mineralisation of filamentous cyanobacteria in Lake Thetis stromatolites, western Australia. Geobiology. 2018;**16**(2):203-215

[12] Holland HD. Volcanic gases, black smokers, and the great oxidation event. Geochimica et Cosmochimica Acta. 2002;**66**(21):3811-3826

[13] Anbar AD, Knoll AH. Proterozoic ocean chemistry and evolution: A bioinorganic bridge? Science. 2002; **297**(5584):1137-1142

[14] Yu Y, Chen Y, Li D, Su J. A transient oxygen increase in the Mesoproterozoic ocean at ~1.44 Ga: Geochemical evidence from the Tieling formation, north China platform. Precambrian Research. 2022;**369**:106527

[15] Planavsky NJ, Slack JF, Cannon WF, O'Connell B, Isson TT, Asael D, et al. Evidence for episodic oxygenation in a weakly redox-buffered deep mid-Proterozoic ocean. Chemical Geology. 2018;**483**:581-594

[16] Zhang K, Zhu X, Wood RA, Shi Y, Gao Z, Poulton SW. Oxygenation of the

Mesoproterozoic ocean and the evolution of complex eukaryotes. Nature Geoscience. 2018;**11**(5):345-350

[17] Kump LR, Pavlov A, Arthur MA. Massive release of hydrogen sulfide to the surface ocean and atmosphere during intervals of oceanic anoxia. Geology. 2005;**33**(5):397-400

[18] Lang X, Shen B, Peng Y, Xiao S, Zhou C, Bao H, et al. Transient marine euxinia at the end of the terminal Cryogenian glaciation. Nature Communications. 2018;**9**(1):3019

[19] Large RR, Hazen RM, Morrison SM, Gregory DD, Steadman JA, Mukherjee I. Evidence that the GOE was a prolonged event with a peak around 1900 Ma. Geosystems and Geoenvironment. 2022;**1**(2):100036

[20] Lyons TW, Reinhard CT, Planavsky NJ. The rise of oxygen in earth's early ocean and atmosphere. Nature. 2014;**506**(7488):307-315

[21] Wille M, Nägler TF, Lehmann B, Schröder S, Kramers JD. Hydrogen sulphide release to surface waters at the Precambrian/Cambrian boundary. Nature. 2008;**453**(7196):767-769

[22] Meyer KM, Kump LR, Ridgwell A. Biogeochemical controls on photic-zone euxinia during the end-Permian mass extinction. Geology. 2008;**36**(9):747-750

[23] Turchyn AV, Schrag DP. Cenozoic evolution of the sulfur cycle: Insight from oxygen isotopes in marine sulfate. Earth and Planetary Science Letters. 2006;**241**(3):763-779

[24] Paytan A, Kastner M, Campbell D, Thiemens MH. Sulfur isotopic composition of cenozoic seawater sulfate. Science. 1998;**282**(5393):1459-1462

[25] Reinhard CT, Planavsky NJ, Robbins LJ, Partin CA, Gill BC, Lalonde SV, et al. Proterozoic ocean redox and biogeochemical stasis. Proceedings of the National Academy of Sciences. 2013;**110**(14):5357-5362

[26] Kushkevych I, Hýžová B, Vítězová M, Rittmann SKR. Microscopic methods for identification of sulfate-reducing bacteria from various habitats. International Journal of Molecular Sciences. 2021;**22**(8):4007

[27] Alqaheem Y. A simulation study for the treatment of Kuwait sour gas by membranes. Heliyon. 2021;7(1):e05953

[28] Clark PD, Hyne JB, David TJ. Some chemistry of organosulphur compound types occurring in heavy oil sands: 2. Influence of pH on the high temperature hydrolysis of tetrahydrothiophene and thiophene. Fuel. 1984;**63**(1):125-128

[29] Marriott RA, Pirzadeh P, Marrugo-Hernandez JJ, Raval S. Hydrogen sulfide formation in oil and gas. Canadian Journal of Chemistry. 2015;**94**(4):406-413

[30] Skjevrak I, Standnes DC, Thomsen US, Xu J, Håland K, Kjølhamar A, et al. Field observations of reservoir souring development and implications for the extended growth zone (EGZ) souring model. Journal of Petroleum Science and Engineering. 2021;**204**:108721

[31] Gittel A. Problems caused by microbes and treatment strategies monitoring and preventing reservoir souring using molecular microbiological methods (MMM). In: Applied Microbiology and Molecular Biology in Oilfield Systems. Dordrecht, Netherlands: Springer; 2011

[32] Gieg LM, Jack TR, Foght JM. Biological souring and mitigation in oil reservoirs. Applied Microbiology and Biotechnology. 2011;**92**:263

[33] Sunde E, Torsvik T. Microbial control of hydrogen sulfide production in oil reservoirs. In: Ollivier B, Magot M, editors. Petroleum Microbiology. USA: ASM Press; 2005. pp. 199-213

[34] Wang Q. Generation mechanism and control measures for H_2S in oil wells, Liaohe oilfield. Petroleum Exploration and Development. 2008; **35**(3):349-354

[35] Ma Q, Yang Z, Zhang L, Lin R, Wang X. Generation of hydrogen sulfide during the thermal enhanced oil recovery process under superheated steam conditions. RSC Advances. 2019; **9**(58):33990-33996

[36] Segers R. Methane production and methane consumption: A review of processes underlying wetland methane fluxes. Biogeochemistry. 1998;**41**(1): 23-51

[37] Ma X, Zheng G, Liang M, Xie D, Martinelli G, Sajjad W, et al. Occurrence and origin of H_2S from volcanic reservoirs in Niudong area of the Santanghu Basin, NW China. Geofluids. 2019;**2019**:1279658

[38] Jiang L, Worden RH, Cai C. Generation of isotopically and compositionally distinct water during thermochemical sulfate reduction (TSR) in carbonate reservoirs: Triassic Feixianguan formation, Sichuan Basin, China. Geochimica et Cosmochimica Acta. 2015;**165**:249-262

[39] Sbodio JI, Snyder SH, Paul BD. Regulators of the transsulfuration pathway. British Journal of Pharmacology. 2019;**176**(4): 583-593

[40] Zhang H-F, Klein Geltink RI, Parker SJ, Sorensen PH. Transsulfuration, minor player or crucial for cysteine homeostasis in cancer. Trends in Cell Biology. 2022;**32**(9):800-814

[41] Bailey TS, Pluth MD. Reactions of isolated persulfides provide insights into the interplay between H_2S and persulfide reactivity. Free Radical Biology & Medicine. 2015;**89**:662-667

[42] Francoleon NE, Carrington SJ, Fukuto JM. The reaction of H_2S with oxidized thiols: Generation of persulfides and implications to H_2S biology. Archives of Biochemistry and Biophysics. 2011;**516**(2):146-153

[43] Heitmann T, Blodau C. Oxidation and incorporation of hydrogen sulfide by dissolved organic matter. Chemical Geology. 2006;**235**(1):12-20

[44] Fashina TB, Adesanwo OO, Adebiyi FM. Influence of humic acid on biodegradation of petroleum hydrocarbons in oil-contaminated soils. Energy Sources, Part A: Recovery, Utilization, and Environmental Effects. 2016;**38**(17):2624-2632

[45] Tejeda-Agredano MC, Mayer P, Ortega-Calvo JJ. The effect of humic acids on biodegradation of polycyclic aromatic hydrocarbons depends on the exposure regime. Environmental Pollution (Barking, Essex: 1987). 2014; **184**:435-442

[46] Park S, Kim KS, Kim JT, Kang D, Sung K. Effects of humic acid on phytodegradation of petroleum hydrocarbons in soil simultaneously contaminated with heavy metals. Journal of Environmental Sciences (China). 2011;**23**(12):2034-2041

[47] Whitby LM, Schnitzer M. Humic and fulvic acids in sediments and soils of agricultural watersheds. Canadian Journal of Soil Science. 1978;**58**(2): 167-178

[48] Einsiedl F, Mayer B, Schäfer T. Evidence for incorporation of H_2S in groundwater fulvic acids from stable isotope ratios and sulfur K-edge X-ray absorption near edge structure spectroscopy. Environmental Science & Technology. 2008;**42**(7): 2439-2444

[49] Zuo Z, Chang J, Lu Z, Wang M, Lin Y, Zheng M, et al. Hydrogen sulfide generation and emission in urban sanitary sewer in China: What factor plays the critical role? Environmental Science: Water Research & Technology. 2019;**5**(5):839-848

[50] Fenouil LA, Towler GP, Lynn S. Removal of H_2S from coal gas using limestone: Kinetic considerations. Industrial & Engineering Chemistry Research. 1994;**33**(2):265-272

[51] Vandecruys F, Brauns E, Engelen W, De Schutter F, Vangrunderbeek J. A H_2S-sensor for on-line measurements in a coal gasification system. Solid State Ionics. 1998;**112**(1):95-102

[52] Sawalha H, Maghalseh M, Qutaina J, Junaidi K, Rene ER. Removal of hydrogen sulfide from biogas using activated carbon synthesized from different locally available biomass wastes - A case study from Palestine. Bioengineered. 2020;**11**(1):607-618

[53] Giampaoli S, Valeriani F, Gianfranceschi G, Vitali M, Delfini M, Festa MR, et al. Hydrogen sulfide in thermal spring waters and its action on bacteria of human origin. Microchemical Journal. 2013;**108**:210-214

[54] Abdollahi M, Hosseini A. Hydrogen sulfide. In: Wexler P, editor. Encyclopedia of Toxicology. 3rd ed. Oxford: Academic Press; 2014. pp. 971-974

[55] Farahbod F, Farahmand S. Introduction of novel process for sweetening of sour crude oil: Optimization of process. Journal of Energy Resources Technology. 2016;**139**(2):022907

[56] Li D-p, Zhang L, Yang J-w, Lu M-x, Ding J-h, Liu M-l. Effect of H_2S concentration on the corrosion behavior of pipeline steel under the coexistence of H_2S and CO_2. International Journal of Minerals, Metallurgy, and Materials. 2014;**21**(4):388-394

[57] Naman SA, Ture IE, Veziroglu TN. Industrial extraction pilot plant for stripping H_2S gas from Black Sea water. International Journal of Hydrogen Energy. 2008;**33**(22):6577-6585

[58] Farahat SA, Kishk NA. Cognitive functions changes among Egyptian sewage network workers. Egyptian Journal of Occupational Medicine. 2009; **33**(2):253-270

[59] Lim E, Mbowe O, Lee ASW, Davis J. Effect of environmental exposure to hydrogen sulfide on central nervous system and respiratory function: A systematic review of human studies. International Journal of Occupational and Environmental Health. 2016;**22**(1): 80-90

[60] Heldal KK, Austigard ÅD, Svendsen KH, Einarsdottir E, Goffeng LO, Sikkeland LI, et al. Endotoxin and hydrogen sulphide exposure and effects on the airways among waste water workers in sewage treatment plants and sewer net system. Annals of Work Exposures and Health. 2019;**63**(4):437-447

[61] Dordević D, Jančíková S, Vítězová M, Kushkevych I. Hydrogen sulfide toxicity in the gut environment: Meta-analysis of sulfate-reducing and lactic acid bacteria in inflammatory processes. Journal of Advanced Research. 2021;**27**:55-69

[62] Barton LL, Fardeau ML, Fauque GD. Hydrogen sulfide: A toxic gas produced by dissimilatory sulfate and sulfur reduction and consumed by microbial oxidation. Metal Ions in Life Sciences. 2014;**14**:237-277

[63] Xue Y, Voordouw G. Control of microbial sulfide production with biocides and nitrate in oil reservoir simulating bioreactors. Frontiers in Microbiology. 2015;**6**:1387

[64] Zuo Z, Song Y, Ren D, Li H, Gao Y, Yuan Z, et al. Control sulfide and methane production in sewers based on free ammonia inactivation. Environment International. 2020;**143**:105928

[65] Zheng M, Zuo Z, Zhang Y, Cui Y, Dong Q, Liu Y, et al. Nitrite production from urine for sulfide control in sewers. Water Research. 2017;**122**: 447-454

[66] Engelbrektson A, Hubbard CG, Tom LM, Boussina A, Jin YT, Wong H, et al. Inhibition of microbial sulfate reduction in a flow-through column system by (per)chlorate treatment. Frontiers in Microbiology. 2014;**5**:315

[67] Nemati M, Jenneman GE, Voordouw G. Mechanistic study of microbial control of hydrogen sulfide production in oil reservoirs. Biotechnology and Bioengineering. 2001; **74**(5):424-434

[68] Hubert C, Voordouw G. Oil field souring control by nitrate-reducing Sulfurospirillum spp. that outcompete sulfate-reducing bacteria for organic electron donors. Applied and Environmental Microbiology. 2007; **73**(8):2644-2652

[69] Zhu X. Microbial control. In: Wang Q, editor. Flow Assurance. Saudi Arabia: Gulf Professional Publishing; 2022. pp. 709-773

[70] Onaizi SA, Gawish MA, Murtaza M, Gomaa I, Tariq Z, Mahmoud M. H_2S scavenging capacity and rheological properties of water-based drilling muds. ACS Omega. 2020;**5**(47):30729-30739

[71] Vo HT, Imai T, Fukushima M, Promnuan K, Suzuki T, Sakuma H, et al. Enhancing the biological oxidation of H_2S in a sewer pipe with highly conductive concrete and electricity-producing bacteria. International Journal of Environmental Research and Public Health. 2023;**20**(2):1459

[72] Khan C, Amin R, Madden G. Effects of CO_2 and acid gas injection on enhanced gas recovery and storage. Journal of Petroleum Exploration and Production Technology. 2013;**3**(1):55-60

[73] Koyanbayev M, Wang L, Wang Y, Hashmet MR. Advances in sour gas injection for enhanced oil recovery-an economical and environmental way for handling excessively produced H2S. Energy Reports. 2022;**8**:15296-15310

[74] Ji Z, Zhang Q, Gao Y, Wang J, He C, Han L, et al. Experimental study on conformance control using acidic nanoparticles in a heterogeneous reservoir by flue gas flooding. Energies. 2023;**16**(1):315

[75] Wang J, Tian S, Liu X, Wang X, Huang Y, Fu Y, et al. Molecular dynamics simulation of the oil–water interface behavior of modified graphene

oxide and its effect on interfacial phenomena. Energies. 2022;**15**(12):4443

[76] Alnili F, Barifcani A. Efficient separation scheme for binary mixture of CO_2 and H_2S using aromatic components. Separation Science and Technology. 2018;**53**(2):312-319

[77] Luo E, Fan Z, Hu Y, Zhao L, Wang J. An evaluation on mechanisms of miscibility development in acid gas injection for volatile oil reservoirs. Oil & Gas Science and Technology - Revue d'IFP Energies nouvelles. 2019;**74**:59

[78] Khan C, Amin R, Madden G. Effects of CO_2 and acid gas injection on enhanced gas recovery and storage. Journal of Petroleum Exploration and Production Technology. 2013;**3**:55-60

[79] Khairulin S, Kerzhentsev M, Salnikov A, Ismagilov ZR. Direct selective oxidation of hydrogen sulfide: Laboratory, pilot and industrial tests. Catalysts. 2021;**11**(9):1109

[80] Zhao W, Manno M, Al Wahedi Y, Tsapatsis M, Stein A. Regenerable sorbent pellets for the removal of dilute H_2S from claus process tail gas. Industrial & Engineering Chemistry Research. 2021;**60**(50):18443-18451

[81] Metz B, Davidson O, de Coninck H, Loos M, Meyer L. Carbon Dioxide Capture and Storage: Special Report of the Intergovernmental Panel on Climate Change. Cambridge, UK: Cambridge University Press; 2005

[82] Chiquet P, Broseta D, Thibeau S. Wettability alteration of caprock minerals by carbon dioxide. Geofluids. 2007;**7**(2):112-122

[83] Iglauer S, Mathew MS, Bresme F. Molecular dynamics computations of brine-CO_2 interfacial tensions and brine-CO_2-quartz contact angles and their effects on structural and residual trapping mechanisms in carbon geo-sequestration. Journal of Colloid and Interface Science. 2012;**386**(1): 405-414

[84] Parmigiani A, Huber C, Bachmann O, Chopard B. Pore-scale mass and reactant transport in multiphase porous media flows. Journal of Fluid Mechanics. 2011;**686**:40-76

[85] Wong S, Keith D, Wichert E, Gunter B, McCann T. Economics of acid gas reinjection: An innovative CO_2 storage opportunity. In: Gale J, Kaya Y, editors. Greenhouse Gas Control Technologies - 6th International Conference. Oxford: Pergamon; 2003. pp. 1661-1664

[86] Bachu S, Haug K, Michael K, Buschkuehle BE, Adams JJ. Deep injection of acid gas in western Canada. In: Tsang C-F, Apps JA, editors. Developments in Water Science. Vol. 52. Netherlands: Elsevier; 2005. pp. 623-635

[87] Lubaś J, Szott W. 15—Year experience of acid gas storage in the natural gas structure of Borzęcin, Poland. In: Second EAGE CO_2 Geological Storage Workshop. Berlin, Germany: European Association of Geoscientists & Engineers; 2010. pp. 333-338

[88] Baroni A et al. Modelling surface displacements associated with CO_2 re-injection at Krechba. In: 45th U.S. Rock Mechanics/Geomechanics Symposium. San Francisco, California: American Rock Mechanics Association (ARMA); 2011

[89] Heinrich J, Herzog H, Reiner D. Environmental assessment of geologic storage of CO2. In: Second Annual Conference on Carbon Sequestration.

Alexandria, Virginia: US Department of Energy; 2003

[90] Vilarrasa V, Rutqvist J. Thermal effects on geologic carbon storage. Earth-Science Reviews. 2017;**165**:245-256

[91] Carroll JJ. Phase equilibria relevant to acid gas injection, part 1-non-aqueous phase behaviour. Journal of Canadian Petroleum Technology. 2002;**41**(6):7

[92] Carroll JJ. Phase equilibria relevant to acid gas injection: Part 2-aqueous phase behaviour. Journal of Canadian Petroleum Technology. 2002; **41**(7):5

[93] Rasool MH, Ahmad M, Ayoub M. Selecting geological formations for CO_2 storage: A comparative rating system. Sustainability. 2023;**15**(8):6599

[94] Hassan Baabbad HK, Artun E, Kulga B. Understanding the controlling factors for CO_2 sequestration in depleted shale reservoirs using data analytics and machine learning. ACS Omega. 2022; 7(24):20845-20859

[95] Newell P, Ilgen AG. Overview of geological carbon storage (GCS). In: Newell P, Ilgen AG, editors. Science of Carbon Storage in Deep Saline Formations: Process Coupling Across Time and Spatial Scales. Amsterdam, Netherlands: Elsevier; 2019. pp. 1-13

[96] Tomić L, Karovic Maricic V, Danilović D, Crnogorac M. Criteria for CO_2 storage in geological formations. Podzemni Radovi. 2018;**2018**:61-74

[97] Callas C, Saltzer S, Davis J, Hashemi S, Kovscek A, Okoroafor E, et al. Criteria and workflow for selecting depleted hydrocarbon reservoirs for carbon storage. Applied Energy. 2022; **324**:119668

[98] Raza A, Rezaee R, Gholami R, Bing C, Nagarajan R, Ali M. A screening criterion for selection of suitable CO_2 storage sites. Journal of Natural Gas Science and Engineering. 2016;**28**: 317-327

[99] Celia MA, Bachu S, Nordbotten JM, Bandilla KW. Status of CO_2 storage in deep saline aquifers with emphasis on modeling approaches and practical simulations. Water Resources Research. 2015;**51**(9):6846-6892

[100] Bourg IC, Beckingham LE, DePaolo DJ. The nanoscale basis of CO_2 trapping for geologic storage. Environmental Science & Technology. 2015;**49**(17):10265-10284

[101] Ajayi T, Gomes JS, Bera A. A review of CO_2 storage in geological formations emphasizing modeling, monitoring and capacity estimation approaches. Petroleum Science. 2019;**16**(5): 1028-1063

[102] Hamza A, Hussein IA, Al-Marri MJ, Mahmoud M, Shawabkeh R, Aparicio S. CO_2 enhanced gas recovery and sequestration in depleted gas reservoirs: A review. Journal of Petroleum Science and Engineering. 2021;**196**:107685

[103] Rodosta TD, Litynski JT, Plasynski SI, Hickman S, Frailey S, Myer L. U.S. department of energy's site screening, site selection, and initial characterization for storage of CO_2 in deep geological formations. Energy Procedia. 2011;**4**:4664-4671

[104] Gupta A. Capacity and constraints for carbon dioxide sequestration in aquifers and depleted oil/gas reservoirs in carbonate environment. In: SPE International Conference on CO_2 Capture, Storage, and Utilization. New Orleans: Society of Petroleum Engineers (SPE); 2010

[105] Hauber G. Norway's Sleipner and Snøhvit CCS: Industry Models or Cautionary Tales? Lakewood, Ohio: Institute for Energy Economics and Financial Analysis (IEEFA); 2023. p. 62

[106] Warnecki M, Wojnicki M, Kuśnierczyk J, Szuflita S. Study of the long term acid gas sequestration process in the Borzęcin structure: Measurements insight. Energies. 2021; **14**(17):5301

[107] Wang H, Zeuschner J, Eremets M, Troyan I, Willams J. Stable solid and aqueous H_2CO_3 from CO_2 and H_2O at high pressure and high temperature. Scientific Reports. 2016;**6**(1):19902

[108] De Silva GPD, Ranjith PG, Perera MSA. Geochemical aspects of CO_2 sequestration in deep saline aquifers: A review. Fuel. 2015;**155**:128-143

[109] Bachu S, Bennion DB. Chromatographic partitioning of impurities contained in a CO_2 stream injected into a deep saline aquifer: Part 1. Effects of gas composition and in situ conditions. International Journal of Greenhouse Gas Control. 2009;**3**(4): 458-467

[110] Knauss KG, Johnson JW, Steefel CI. Evaluation of the impact of CO_2, co-contaminant gas, aqueous fluid and reservoir rock interactions on the geologic sequestration of CO_2. Chemical Geology. 2005;**217**(3):339-350

[111] Xu T, Apps JA, Pruess K, Yamamoto H. Numerical modeling of injection and mineral trapping of CO_2 with H_2S and SO_2 in a sandstone formation. Chemical Geology. 2007; **242**(3):319-346

[112] Schaef HT, McGrail BP, Owen AT, Arey BW. Mineralization of basalts in the CO_2–H_2O–H_2S system. International Journal of Greenhouse Gas Control. 2013;**16**:187-196

[113] Ofori K, Phan CM, Barifcani A, Iglauer S. An investigation of some H_2S thermodynamical properties at the water interface under pressurised conditions through molecular dynamics. Molecular Physics. 2022;**120**(6):e2011972

[114] Ofori K, Phan CM, Barifcani A, Iglauer S. Some interfacial properties of water and CO_2/H_2S at quasireservoir conditions: A molecular dynamics study. SPE Journal. 2023;**28**(2):783-795

[115] Li Q et al. Effects of acid gas reinjection on enhanced natural gas recovery and carbon dioxide geological storage: Investigation of the right bank of the Amu Darya River. In: Wu Y, Carroll JJ, Zhu W, editors. Carbon Dioxide Capture and Acid Gas Injection. New York: John Wiley & Sons; 2017. pp. 221-240

[116] Bachu S, Gunter WD. Acid-gas injection in the Alberta basin, Canada: A CO_2-storage experience. Geological Society, London, Special Publications. 2004;**233**(1):225-234

[117] Bachu S, Nordbotten JM, Celia MA. Evaluation of the spread of acid-gas plumes injected in deep saline aquifers in western Canada as an analogue for CO2 injection into continental sedimentary basins. In: Rubin ES et al., editors. Greenhouse Gas Control Technologies 7. Oxford: Elsevier Science Ltd; 2005. pp. 479-487

[118] Bachu S, Carroll JJ. In-situ phase and thermodynamic properties of resident brine and acid gases (CO_2 and H_2S) injected into geological formations. In: Greenhouse Gas Control Technologies 7. Vancouver, Canada: Elsevier Science Ltd; 2005

[119] Mi J, Zhang B, Shen Z, Huang W, Casalins A, Liu C. The experimental study on H_2S generation during thermal recovery process for heavy oil from the eastern Venezuela Basin. Journal of Natural Gas Geoscience. 2017;**2**(3): 201-208

[120] Gunnarsson I et al. Injection of H_2S from Hellisheiði power plant, Iceland. In: Thirty-Sixth Workshop on Geothermal Reservoir Engineering. Stanford, California: Stanford Geothermal Program; 2011

[121] Georgiadis AG, Charisiou N, Yentekakis IV, Goula MA. Hydrogen sulfide (H_2S) removal via MOFs. Materials. 2020;**13**(16):3640

[122] Gutierrez-Sevillano JJ, Martin-Calvo A, Dubbeldam D, Calero S, Hamad S. Adsorption of hydrogen sulphide on metal-organic frameworks. RSC Advances. 2013;**3**(34): 14737-14749

[123] Koschel D, Coxam J-Y, Majer V. Enthalpy and solubility data of H_2S in water at conditions of interest for geological sequestration. Industrial & Engineering Chemistry Research. 2007; **46**(4):1421-1430

[124] Koschel D, Coxam J-Y, Majer V. Enthalpy and solubility data of H_2S in aqueous salt solutions at conditions of interest for geological sequestration. Industrial & Engineering Chemistry Research. 2013;**52**(40): 14483-14491

[125] Chapoy A, Mohammadi AH, Tohidi B, Valtz A, Richon D. Experimental measurement and phase behavior modeling of hydrogen sulfide—Water binary system. Industrial & Engineering Chemistry Research. 2005;**44**(19):7567-7574

[126] Springer RD, Wang P, Anderko A. Modeling the properties of H_2S/CO_2/salt/water systems in wide ranges of temperature and pressure. SPE Journal. 2015;**20**(5):1120–1134

[127] Machel HG. Geological and hydrogeological evaluation of the Nisku Q-Pool in Alberta, Canada, for H_2S and/or CO_2 storage. Oil & Gas Science and Technology - Revue d'IFP Energies nouvelles. 2005;**60**(1):51-65

[128] Ng H-J, Carroll JJ, Maddocks J. Impact of thermophysical properties research on acid-gas injection process design. In: Gas Processors Association Annual Convention. Nashville, TN: Gas Processors Association; 1999